Electromagnetic Waves Propagation in Complex Matter

Electromagnetic Waves Propagation in Complex Matter

Editor

Gustaf Kron

scitus
academics

Electromagnetic Waves Propagation in Complex Matter
Edited by **Gustaf Kron**

ISBN: 978-1-68117-292-7
Library of Congress Control Number: 2016937475

SCITUS Academics LLC,
www.scitusacademics.com
Box No. 4766, 616 Corporate Way,
Suite 2, Valley Cottage,
NY 10989

Notice

Printed in United States of America on Acid Free Paper

Preface

Electromagneticwaves are waves which can travel through the vacuum of outer space. Mechanical waves, unlike electromagnetic waves, require the presence of a material medium in order to transport their energy from one location to another. Light, microwaves, x-rays, and TV and radio transmissions are all kinds of electromagnetic waves. They are all the same kind of wavy disturbance that repeats itself over a distance called the wavelength. The mechanism of energy transport through a medium involves the absorption and reemission of the wave energy by the atoms of the material. When an electromagnetic wave impinges upon the atoms of a material, the energy of that wave is absorbed. The absorption of energy causes the electrons within the atoms to undergo vibrations. After a short period of vibrational motion, the vibrating electrons create a new electromagnetic wave with the same frequency as the first electromagnetic wave. While these vibrations occur for only a very short time, they delay the motion of the wave through the medium. Once the energy of the electromagnetic wave is reemitted by an atom, it travels through a small region of space between atoms. Once it reaches the next atom, the electromagnetic wave is absorbed, transformed into electron vibrations and then reemitted as an electromagnetic wave. Electromagnetic Waves Propagation in Complex Matter emphasizes the topics of wave propagation and interaction with matters. The book bridges the gap between physics and engineering in these issues.

Table of Contents

1

CHAPTER

ELECTROMAGNETIC WAVES IN CONTAMINATED SOILS

Arvin Farid[1], Akram N. Alshawabkeh[2] and Carey M. Rappaport[3]

[1] Boise State University, USA

1. INTRODUCTION

Soil is a complex, potentially heterogeneous, lossy, and dispersive medium. Modeling the propagation and scattering of electromagnetic (EM) waves in soil is, hence, more challenging than in air or in other less complex media. This chapter will explain fundamentals of the numerical modeling of EM wave propagation and scattering in soil through solving Maxwell's equations using a finite difference time domain (FDTD) method. The chapter will explain how: (i) the lossy and dispersive soil medium (in both dry and fully water-saturated conditions), (ii) a fourth phase (anomaly), (iii) two different types of transmitting antennae (a monopole and a dipole), and (iv) required absorbing boundary conditions can numerically be modeled. This is described through two examples that simulate the detection of DNAPL (dense nonaqueous-phase liquid) contamination in soil using Cross-well radar (CWR). CWR —otherwise known as cross-borehole GPR (ground penetrating radar)—modality was selected to eliminate the need for simulation of the roughness of the soil-air interface. The two examples demonstrate the scattering effect of a dielectric anomaly (representing a DNAPL pool) on the EM wave propagation through soil.

The objective behind selecting these two examples is twofold: (i) explanation of the details and challenges of numerical modeling of EM wave propagation and scattering through soil for an actual problem (in this case,

DNAPL detection), and (ii) demonstration of the feasibility of using EM waves for this actual detection problem.

In addition, the results of an experimental simulation of Example 1 (the case with the monopole antenna) will be analyzed and discussed. The results of the corresponding numerically simulated example will then be compared and validated against the above-mentioned experimental results. A conclusion section will close the chapter.

Before explaining the numerical modeling and its challenges, some background about DNAPLs and their detection technologies, including the CWR method, is explained in the following section.

2. BACKGROUND

Cross-well radar (CWR), otherwise known as cross-borehole ground-penetrating radar (cross-borehole GPR) is a minimally invasive method that uses high frequency electromagnetic (EM) waves transmitted and received by antennae in the subsurface to image objects of contrasting dielectric properties. In order to assess the feasibility of using CWR for detection of dense nonaqueous-phase liquid (DNAPL) pools in deeper layers, this technique should be numerically simulated and evaluated. CWR is more appropriate for deep investigations. For near-surface sensing such as bridge-deck health monitoring, methods such as GPR are more appropriate (Belli et al., 2009 and 2009a). GPR studies require addressing the scattering due to the rough soil-air interface in the forward model as well as through the inversion process (Firoozabadi et al., 2007).This chapter describes a numerical modeling approach to Maxwell's equations using a finite difference time domain (FDTD) solution with both monopole and dipole antennae to simulate the scattering and propagation of EM waves in DNAPL-contaminated media. An FDTD code, originally developed for detection of mines using two-dimensional (2D) surface-reflection ground-penetrating radar (GPR) through non-dispersive media, was revised and upgraded for three-dimensional (3D) cross-borehole wave propagation in heterogeneous soils. The three dimensional FDTD code was enhanced to accommodate dispersive media and was used to model EM wave propagation in heterogeneous soils. This chapter describes the effect of the radiation patterns of two different antennae on propagation and scattering of EM waves through the soil subsurface and its potential for detection of DNAPL pools. In order to evaluate the feasibility of using the CWR method to detect DNAPL pools, illustrative examples with and without the presence of the DNAPL pool were analyzed. The results show considerable diagnostic potential to detect contaminated zones with DNAPLs using EM waves through CWR.

DNAPLs are separate-phase hydrocarbon liquids denser than water, such as chlorinated solvents (tetrachloroethene (PCE) and trichloroethylene (TCE)), wood preservatives, coal tar wastes, and pesticides. DNAPLs may not usually be found as a free phase in soil cores or accumulated in monitoring wells. Based on this lack of observable evidence of DNAPL pools, investigators may conclude that no DNAPL is present, when it may be present in substantial quantities at residual saturation as large as 70% to 80% of the total porosity (ITRC, 2000).

There are varieties of invasive techniques to detect DNAPLs such as direct push probe techniques (e.g., direct soil sampling or indirect sampling such as negative ion sensor) and use of in situ tracers (e.g., PITT, or partitioning interwell tracer test), excavating test pits, and groundwater profiling. Most invasive techniques just provide point-sources of information and may help the DNAPL pool to spread through the substantial number of required boreholes and excavated trenches, while noninvasive geophysical techniques noninvasively or minimally invasively detect DNAPLs. They can use different types of waves (e.g., electromagnetic, acoustic, etc.) to indirectly reconstruct images to characterize or detect anomalies as DNAPLs within heterogeneous media. These methods are minimally invasive techniques that discriminate the contrast between local physical properties of the background and target to produce images of the subsurface.

Cross-well P-wave transmission at 90 kHz was used in a sand pack before and after introducing NAPLs by Geller et al. (2000). The results indicated that small NAPL saturations may be more easily detected with amplitude than with travel time data, but the relationships between the amplitude changes and NAPL saturation may be more complex than those for velocity.

Cross-hole complex resistivity was also applied to a contaminated vadose zone by Grimm and Olhoeft (2004) to predict the general distribution of DNAPLs at parts per thousand concentrations, specifically widespread near-surface contamination and in the vadose zone immediately underneath the source.

Complex resistivity (CR) is a technique that so far has shown a potential in detecting DNAPL pools (Blackhawk Geoservices Inc., 2008).

Smith-Rose (1993& 1935) is among the earliest scientists who studied the dielectric permittivity of different soils. Some physical properties of a typical DNAPL (i.e., PCE) are compared with those of different soils, water, and air in Table 1.

As seen, dielectric permittivity and effective (= DC + alternating field) electrical conductivity of PCE are lower than those of water and relatively similar to air or some dry soils. Thus, a PCE-contaminated region of the saturated zone should provide a contrast with a similar magnitude of the unsaturated-saturated zone interface (water table). The values of dielectric permittivity for water and soil are dispersive (depend on frequency), but most DNAPLs are nonpolar molecules with minimal frequency dependence and therefore with almost no dispersive characteristics. In frequencies greater than 10 MHz and less than 1 GHz, the dielectric permittivity is controlled by the polarization of individual water molecules and is therefore dependent on moisture content (Binley et al., 2001). Dielectric permittivity in the lower MHz range of frequency (< 10 MHz) depends on particle shape and mineralogy, electrolyte type and concentration, particle orientation, and soil electrolyte interaction (Rinaldi & Francisca, 1999). In the upper MHz range of frequency (10 to 1000 MHz), the real part of dielectric permittivity is affected by the polar contribution of bound and free water molecules (Hoekstra & Doyle, 1971; Hoekstra & Delaney, 1974; Selig & Mansukhani, 1975; Dobson et al., 1985; Hallikainen et al., 1985; and Arulanandan, 1964). Sachs and Spiegler (1964) presented a model for dielectric permittivity of soil mixtures at different frequencies. This model is based on an equivalent circuit for conductive particles in electrolytes and fitted in the dielectric plane by adjustment of three parameters. Arulanandan and Smith (1973) explained some aspects of dielectric dispersion based on the Sachs and Spiegler (1964) model in a frequency range from 10^6 to 10^8 Hz. Another model that explains the dispersion of the soil relative dielectric permittivity, otherwise known as dielectric constant, was presented by Thevanayagham (1995), which considers the effect of particle orientation relative to the electric field. Weedon and Rappaport (1997) and Rappaport et al. (1999) presented a model to predict the dispersive nature of the dielectric permittivity and electrical conductivity of soils, using a rational Z-transform approximation function of the conductivity. This technique uses the values of dielectric permittivity and electrical conductivity of the bulk mixture, and does not deal with the components and their volumetric content. It is basically a technique to convert the bulk values for the matrix to a form applicable to the FDTD method to take the dispersive nature of the resultant mixture into account.

Table 1. Physical Properties of PCE, Water and Soil at 25°C (Degrees Celsius)(Brewster & Annan, 1994), (Von Hippel, 1953), (Hipp, 1974), (Weedon & Rappaport, 1997), (Rappaport et al., 1999).

Material	Density ρ (g/cm3)	Real Part of Dielectric Constant* (ε_r)		**** Effective Electrical Conductivity σ_e (1/Ω.m)	
		f = 100 MHz	f = 1.5 GHz	f = 100 MHz	f = 1.5 GHz
PCE	1.62	2.28	2.28	2.63×10^{-6}	1.52×10^{-4}
Water**	1.00	78	77.14	2.2×10^{-3}	0.509
Dry Sand	1.30	2.55	2.55	1.8×10^{-4}	2.1×10^{-3}
Soil***	1.60	6.912	6.50	0.1329	0.1

* Dielectric constant is also known as relative dielectric permittivity
** Pure water
*** 20% moist Bosnian clay loam
**** Effective electrical conductivity (σ_e) = Static or DC conductivity (σ_s) + Alternating field conductivity (σ_a)

The loss (attenuation) in soil is a function of a variety of factors such as soil type and mineralogy, moisture content, electrical conductivity, and frequency. Higher frequency EM waves attenuate faster. In other words, soil acts as a low-pass filter. However, higher frequency EM waves provide higher image resolution. There is a tradeoff between the penetration depth and image resolution. Therefore, feasibility of the CWR to detect an average size DNAPL pool using proper frequency range is the first step to detection, which is the main goal of this chapter.

Soil, water, and DNAPL mixed at DNAPL saturations of 0% to 50% (i.e., water saturation of 50% to 100% below water table; i.e., no air) have been evaluated by Ajo-Franklin et al. (2004) to have a dielectric constant of 9 (50% water saturation soil with 50% DNAPL saturation) to 24 (fully (100%) water-saturated soil with 0% DNAPL saturation). These values are different from the dielectric constant of pure DNAPL ($\varepsilon' \approx 2.3$) and vary based on porosity and degree of DNAPL saturation as the result of mixture.

There are some expected limitations to using CWR to detect DNAPLs in soil and groundwater such as: (a) use of CWR to detect DNAPLS may not perform as well in dry or partially saturated zones as in water saturated zones, due to the weaker contrast in electromagnetic properties of DNAPLs and dry soils compared to the one between DNAPLs and water saturated soils, (b) the concentration (or contaminant saturation) must be fairly high, which makes the detection of DNAPLs using CWR often limited to identifying DNAPL pool sources (high DNAPL saturation) and not plumes, and (c) the spacing between the wells strongly influences the effectiveness of CWR (as the separation between transmitting and receiving antennae increases, the radar wave amplitude attenuates, which creates greater difficulty in distinguishing the wave from background noises) (ITRC, 2000).

Therefore, in this chapter a simple case of DNAPL dominantly replacing water in the pores (DNAPL saturation of 80% to 100%) is simulated numerically. Based on Bruggeman-Hanai-Sen (BHS) model (Sen et al., 1981), due to low dielectric constant of soil grains (≈ 5, (Ajo-Franklin et al., 2004)), the bulk dielectric constant of the soil mixture within the DNAPL pool (close to 100% DNAPL saturation) values between 2.3 and 5. Hence, for comparison, acrylic was selected as the dielectric object of ′ ≈ 2.6 (Weast, 1974) since it has been used in another experimental work by the authors (Farid et al., 2006). This strong contrast in dielectric constant between the dielectric constant of the acrylic (≈ 2.6) and water-saturated soils (≈ 15 - 25) makes detecting the scatterer in the saturated soil using radar-based geophysical methods feasible.

The U.S. Geological Survey used 500-MHz surface GPR among other methods to monitor the location and migration of the subsequent plume (Sneddon et al., 2000).

Bradford and Wu (2007) tested 3D multi-fold GPR on a small scale controlled DNAPL release to detect contaminated soil with demonstrated success.

While GPR is the least invasive radar-based method, it is not a practical one for deep investigations. Cross-Well Radar (CWR), otherwise known as cross-borehole GPR, may be a more effective method for deep contaminant detection. CWR is a method that uses EM wave transmission measurement across borehole antennae as opposed to reflection measurements used in surface-reflection imaging methods, such as GPR. CWR overcomes some of depth limitations of surface-reflection imaging GPR. CWR uses antennae that are lowered into sampling wells with cables. Radar waves are emitted from a transmitting antenna in one well and propagate through the ground to a receiving antenna in a second well. The subsurface geology and pore fluids scatter, refract, or reflect the waves. DNAPL pools also provide dielectric permittivity and electrical conductivity contrasts within relative moist soils, which in turn, cause refractions or reflections on EM waves. These scatterings and reflections make DNAPLs amenable to detection by radar. Other than the magnitude of the scattering field, phase, and hence, travel time can also be used for other techniques such as travel time tomography to reconstruct images of the background medium and anomalies within the medium.

Success of radar-based methods as well as any other detection technique depends on the strength of the contrast in properties of the background and scatterer (e.g., DNAPLs, moisture). Dielectric permittivity and electrical conductivity are the main parameters controlling such differences for radar-based techniques.

CWR relies on a one-way travel time whereas the surface-reflection imaging GPR relies on the 2-way travel time. Therefore, the effective antenna separation of CWR can be twice as far as the penetration depth of surface-reflection GPR, theoretically. GPR antennae are designed to be relatively impedance-matched to the ground, and the reflection coefficient at the air-soil interface is close to unity leading to the potential for strong internal reflection. However, there is still the disadvantage of reflections at the air-soil interface, which reduces the amount of received waves reflected by the underground objects. The performance of detection using CWR is highly dependent on seasonal conditions, in particular soil moisture (Daniels et al., 1992). Soil mineral composition and physicochemical properties also influence radar signals. For example, clays tend to attenuate these signals and therefore may limit the skin depth and effectiveness of the method (Anderson & Peltola, 1996). Considering these limiting factors, feasibility of the technique for DNAPL detection should be evaluated by simulating implementation of ultra-wideband CWR. Further understanding of the behavior of 3D EM waves in soil is necessary for the future implementation and feasibility evaluation of CWR for detecting DNAPLs. To achieve this goal, the detection technique using monopole and dipole antennae is modeled numerically in the time domain using an FDTD technique. A critical issue addressed in this research, by direct simulation of the antennae, is the wave interaction at the antenna/soil interface. The technique models the transmitting antenna but measures the electrical field at all grid points on any desired cross-sectional or depth slice (instead of modeling receiving antennae). This is equivalent to having hundreds of receiving antennae in soil, which is not practical. In practical techniques such as cross-well tomography, few antennae are installed and used alternatively as transmitters or receivers to collect data in a multiple-depth, multiple-location manner, and the outcome is used for inversion and image processing techniques (the authors are working on different aspects of cross-well tomography and results will be published in the future). Obviously, the simulation technique presented in this chapter is used only for preliminary evaluation of the feasibility of detecting DNAPL pools using CWR. The outcome can also be used as a forward modeling for future inversion techniques.

Besides, another goal of this chapter is to evaluate the feasibility of DNAPL detection and to investigate if there are points with strong signature by the object on the scattered field, not to discuss inverse modeling. Therefore, receiving antennae are not modeled. The results model no noise. Nevertheless, comparison between the values of the scattered field (by the target) with those of the incident field will provide helpful information that can lead to the critical noise/signal ratio, even without applying any noise. This is discussed in more detail in the following sections. Experimental works are being conducted to evaluate typical levels of noise, compare them to the levels derived from this technique, then apply the noise to the simulation and reevaluate the feasibility (Farid et al., 2006).

3. FTDT MODELING OF MAXWELL'S EQUATIONS

Finite Difference Time Domain (FDTD) is a relatively powerful and very popular method because of its simplicity and despite its instability problems. Therefore, a 3D FDTD code was developed and used to simulate the antenna performance, coupling, and interactions in soils. The FDTD technique was originally introduced by Yee (1966) and is based on time and spatial discretization of Maxwell's equations to obtain solutions for the EM field in the time domain. The technique is numerically implemented by continuously sampling the electromagnetic field over the wave propagation medium discretized into a grid. The differential form of Maxwell's equations in the time domain is as follows (Grant & Philips, 1990; Sheriff, 1989; and Balanis, 1989).

$$\nabla \times \underline{E} = -\frac{\partial \underline{B}}{\partial t} - \underline{J}_m \tag{1}$$

$$\nabla \times \underline{H} = \frac{\partial \underline{D}}{\partial t} + \underline{J}_e \tag{2}$$

$$\nabla \cdot \underline{D} = \rho_e \tag{3}$$

$$\nabla \cdot \underline{B} = \rho_m \tag{4}$$

along with the constitutive relations,

$$\underline{B} = \mu \underline{H} \tag{5}$$

$$\underline{D} = \varepsilon \underline{E} \tag{6}$$

$$\underline{J_e} = \sigma_e \underline{E} \tag{7}$$

$$\underline{J_m} = \sigma_m \underline{H} \tag{8}$$

where E is the electric field density, H is the magnetic field intensity, D is the electric displacement or electric flux density, B is the magnetic flux density, $\underline{J_e}$ is the impressed (source) electric current density, *Jm---* is the impressed (source) magnetic conductive current density, $_e$ is the electrical conductivity, $_m$ is the magnetic resistivity, ε is the dielectric permittivity, is the magnetic permeability, ρ_e is the electric charge density, and ρ_m is the (monopole) magnetic charge density. All of the field parameters, E, H, B, D $\underline{J_e}$,and $\underline{J_m}$ are assumed to be functions of position and time t, while material parameters , , and are functions of position and may be dispersive functions of frequency (and thus, through the Fourier Transform, also of time). Note that the usual multiplicative constitutive relations are written as convolutions for dispersive soil media. Equations (1) through (4) can be also written in integral forms. These four partial differential equations are discretized in a 3D rectangular grid, using a central-difference approximation of two consecutive values for the field components in both space and time. Due to the nature of electric and magnetic fields (e.g., coil loops), these two cannot be located at the same spatial location. In the discretization process, electric and magnetic field components are assigned to the edges of complementary interlocking cubical meshes. Thus, half-space indices are introduced for different field components. Besides, half time steps are introduced to perform the finite difference computation of E based on H and vice versa. Spatial and time steps are respectively represented by the lower indices (i, j, k) and the upper index (n). Some details of the simulation process are explained in the following. Fig. 1 helps to better understand this staggered time and space grid. As seen in the superscripts in the figure, there is ½ time step difference between E and H. Due to the central difference approximation technique in time, H is present at both the $(n + ½)^{th}$ and $(n - ½)^{th}$ time steps in the equations along with E only at the nth time step.

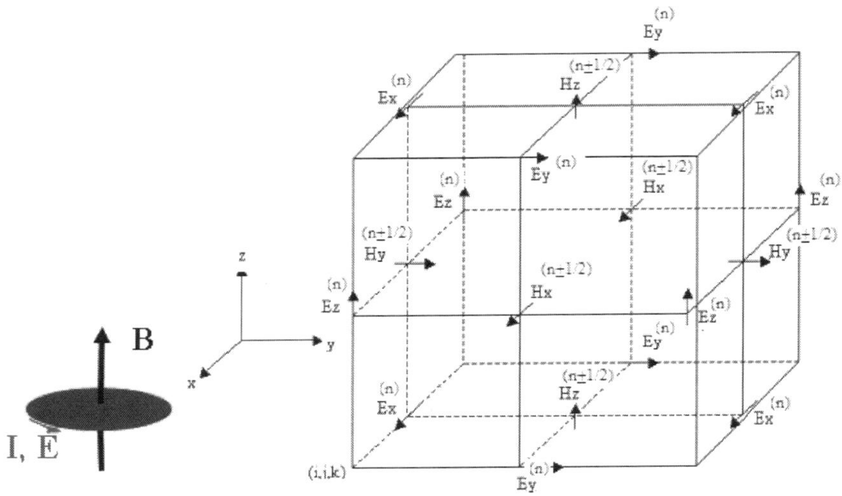

Figure 1. interlocking FDTD cell (Yee cell, 1966).

3.1. Advantages and limitations

The most advantageous characteristic of the FDTD method is its simplicity. Solving Maxwell's equations using FDTD is a simple iterative procedure. However, there are several limitations to the numerical implementation of the FDTD technique through the above-mentioned difference equations. These limitations have been and are subject of research. Some of these challenging limitations impose restrictions on the grid size and time step increments that affect stability and accuracy of the method (Kunz & Luebbers, 1993).

One fundamental restriction requires the longest side of the grid cell to be much shorter than the shortest wavelength of the wave within the cell. A very common restriction assumed in practice is / 10, where is the shortest appreciable wavelength in the excitation signal. This constraint is imposed by both sampling limitation and grid dispersion errors (Kosmas, 2006).

The second restriction develops from the scale and geometry of the problem. The method uses a uniform grid to model small antennae along with a large soil medium with or without contrasting objects. Accordingly, the geometry imposes challenging limitations, especially in computation cost. For example, in the case of a DNAPL contaminated site, dimensions of the background in horizontal directions may be of the order of tens of meters by tens of meters in the XY-plane, while the antenna thickness may be of the order of millimeters to centimeters. This requires solving a large uniform grid with a small grid-cell size in the X and Y directions. At the same time, a small vertical (Z-direction) cell size is required to account for thin DNAPL pools. One way to solve this problem is by using a non-uniform grid, which adds more difficulties to the task of satisfying stability conditions (e.g., Courant condition).

The third limitation is the time-step restriction required to satisfy the Courant condition. The Courant condition restricts time increments to a range in which waves do not travel too far in each time increment, or:

$$\frac{1}{V_{p,\max}\Delta t} \geq \left(\sum_{m=1}^{N} \frac{1}{\Delta x_m^{\ 2}} \right)^{1/2} \tag{9}$$

where N is the grid dimension, x_m is the grid cell size in the mth direction, and $V_{p,max}$ is the maximum wave phase velocity within the model. In a 3D case, Equation (9) will be written in the following form:

$$\frac{1}{V_{p,\max}\Delta t} \geq \left(\frac{1}{\Delta x^2} + \frac{1}{\Delta y^2} + \frac{1}{\Delta z^2} \right)^{1/2} \tag{10}$$

In the case of N-dimensional isotropic cells, Equation (9) can be simplified to $V_{p,\max}\Delta t / \delta \leq 1 / \sqrt{N}$ (e.g., in a three dimensional case, N = 3, and x = y = z =), where the left hand side is called the Courant number. Different cases have demonstrated that using smaller values of t does not necessarily improve the results. However, smaller values for the Courant number may sometimes yield satisfactory results (Kosmas, 2006).

4. MODELING PROCEDURE

4.1. Modeling dispersive nature of soil

Soil is a complex medium to model because of its heterogeneous, lossy, and dispersive nature. Accurate computation of the behavior of soil over a wideband of frequency requires either several individual frequency domain calculations or a robust deconvolution of E(t) from D(t) in the time domain (Rappaport et al., 1999). The convolutional approach to modeling dispersion in soil approximates the frequency domain dispersive complex dielectric constant with rational functions of jω and multiplies the constitutive relations by the denominator. Then, the results are inverse-Fourier-transformed. This model is called the Debye or Lorentz model (Kashiwa & Fukai, 1990; and Gandhi, 1993). Weedon and Rappaport (1997) and Rappaport and Winton (1997) simplified the problem by modeling the conductivity as a simple rational function of Z-transform. Modeling conductivity in terms of the Z-transform variable (Z^{-1} readily transforms to "time delays") simplifies the process of conversion of the generalized dispersive Ohm's Law (J(Z) = σ(Z) E(Z)) to the time domain. To keep the simplicity of the FDTD method, Z-transform function of conductivity in the time domain is given as (Weedon & Rappaport, 1997; and Rappaport et al., 1999):

$$\sigma(Z) = \frac{\underline{\mathrm{J}(Z)}}{\underline{\mathrm{E}(Z)}} = \frac{b_0 + b_1 Z^{-1} + b_2 Z^{-2}}{1 + a_1 Z^{-1}} \tag{11}$$

Using this function, both real and imaginary components of conductivity depend on the sampling interval Δt and the coefficients of the rational Z-transform function (a_1, b_0, b_1, and b_2). The real part of conductivity corresponds to measured conductivity, and the imaginary part corresponds to real dielectric permittivity. Ampere's law can be written in the Z-domain as follows.

$$\nabla \times \underline{\mathrm{H}(Z)} = \varepsilon_{av} \frac{1 - Z^{-1}}{\Delta t} \underline{\mathrm{E}(Z)} + \frac{(1 + Z^{-1})}{2} \sigma(Z) \underline{\mathrm{E}(Z)} \tag{12}$$

4.2. Absorbing boundary conditions

As in simulation of any diffusion application (e.g., heat transfer or water flow through porous media), wave propagation through infinite media is practically modeled in a finite grid. To model the infinity of the flow or wave propagation through the required infinite media by the finite number of grid cells, appropriate boundary conditions are required. The lattice termination absorbing boundary condition used in the FDTD code was based on the second-order one-way wave propagation equation of Mur (1981) and adjusted for lossy soils by Talbot and Rappaport (2000) using the rational function approximation of Equation (11).

4.3. Antenna and excitation

There are some factors that categorize the excitation, such as the source geometry, excitation signal type, and hard or soft source of propagation.

Hard and Soft Sources: The code uses soft sources for all different antenna types. A hard source specifies the total field at the excitation point. These types of sources are usually avoided as they cause undesirable reflections at the source point. Alternatively, a soft source specifies the additional field supplied above the existing background field at the source point. The soft source can be specified at points on a radiating aperture, while hard sources are used for current sources flowing along metal structures. In 2D cases, soft sources propagate well, but extensive testing of 3D FDTD cases has shown difficulties involved in propagation from a soft source unless the physical body of the antennae is modeled. This paper uses soft sources for both monopole and dipole antenna cases of the following sections.

Excitation Signal: A cosine-modulated Gaussian signal (Equation (13)) was used to excite both antennae because of its simplicity and frequency content

$$E^n(i,j,k) = E_0 \exp\left[-\left(\frac{t-t_0}{W}\right)^2\right]\cos(2\pi ft+\phi) \tag{13}$$

where E_0 is the amplitude of the pulse, t is any arbitrary time instant, and t0 is the time instant corresponding to the peak of the Gaussian. Ideally, a Gaussian pulse has infinite duration. However, in numerical implementation it ought to be truncated. Thus, W is chosen as the duration of the pulse. Parameters t_0 and W can be chosen, but not absolutely arbitrarily. In this chapter, W was chosen based on the bandwidth used in a scaled experiment by the authors (Farid et al., 2006). Inappropriate choices of these parameters may introduce instability and noise into the FDTD computation. Starting and truncation of the Gaussian signal should be sufficiently smooth (i.e., the field value at the truncation time should be sufficiently small) to alleviate introduction of frequencies with spectrum levels above the one that a single or double precision calculation can tolerate. Large spectrum energy of the incident field may cause instability and noise, which in turn, results in computation corruption. This is true for any type of pulse. For the Gaussian type pulse, experience has shown that a choice of $t_0 \geq 4W$ ensures smooth truncation, and in turn, good performance (Kosmas, 2006).

4.4. Geometry of the source

To model the DNAPL detection and evaluate its feasibility, two different antenna types were modeled, a monopole and a dipole. Some fundamental parameters and facts about these simulations are explained in the following.

- The monopole antenna is simply a coaxial cable, schematically shown in Fig. 2(a) with its shield removed at a specific length from the tip (referred to as extended dielectric length). In a simple monopole antenna, the most efficient radiation occurs when the exposed dielectric and center conductor length is ¼ of the excitation wavelength. For the range of frequency of interest of this and other ongoing and further research (400-2200 MHz), and based on the dielectric permittivity of the dielectric part of the antennae (≈ 2.1), this length should be between 2.35 cm and 12.94 cm. Different types of antennae have different radiation patterns. The radiation pattern is a visual way of representing how an antenna distributes energy into the surrounding space. Different methods (parameters) can be used to demonstrate the wave propagation through space. One popular method is to use power as the key parameter. This method visualizes how and in which direction power is concentrated by the antenna to propagate away. This is possible by drawing contours of equal power. Other parameters, such as field magnitude or phase, can be used for visualization purposes. Contours of these parameters are drawn to visualize radiation patterns. A monopole antenna radiates waves in all different directions.

A dipole antenna consists of two straight metallic or dielectric parts connected at the center to a feed line. A typical dipole antenna is shown in

Fig. 2(b). This antenna type constitutes the main RF radiating and receiving element in various sophisticated types. Dipoles are inherently balanced antennae due to bilateral symmetry. The best dipole length is usually ½ wavelengths.

5. EVALUATING FEASIBILITY OF DNAPL DETECTION

As mentioned before, two different antennae are simulated in the code to model the problem and evaluate the feasibility of the use of the CWR method to detect dielectric objects in soil. The first case models a small-scale modeled monopole antenna as a soft source of the wave within a fully water-saturated sandy soil medium (degree of water-saturation = S_r = 1) and the gravimetric moisture content = w = 17%). The second case is the simulation of a dipole antenna in a larger size medium of the same fully saturated sandy soil (S_r = 1 and w = 17%). This value of moisture content is selected to model a soil medium similar to the existing fully saturated sandy soil in a pilot-scale facility constructed at Northeastern University (referred to as SoilBED) for subsurface sensing and imaging experimentation. This 17% moisture content is the reference moisture content for much further experimental research by the authors (for more information refer to Farid et al., 2006) in the SoilBED facility. Hydrological modeling of DNAPLs and the dielectric properties converted through a generic petrophysical relationship in the simulation is desired but does not fit in the extent of this chapter. Details of the two above-mentioned simulated cases follow.

5.1. Monopole antenna case (A)

The pilot-scale simulation of the SoilBED facility (Farid et al., 2006) is explained in this section. In the first case, a 5 mm-thick monopole antenna is modeled within a fully saturated sandy soil background. The size of the medium under study was selected to satisfy limitations of the FDTD code as well as the experiment. Table 2(a) summarizes details about the geometry and grid size of the soil medium. The simulation is driven by a cosine modulated Gaussian time pulse at a reasonably high frequency (1.5 GHz).

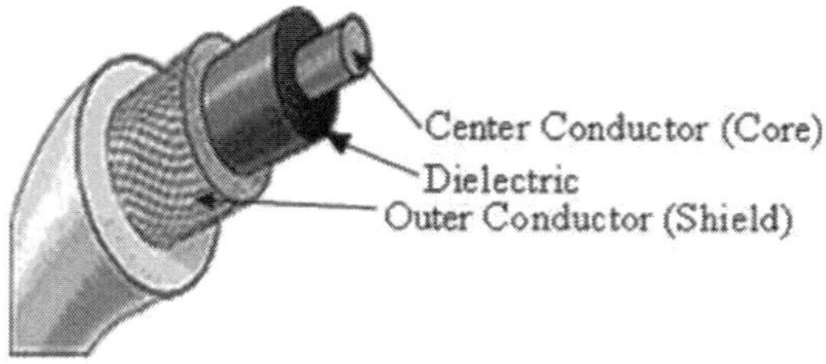

(a)

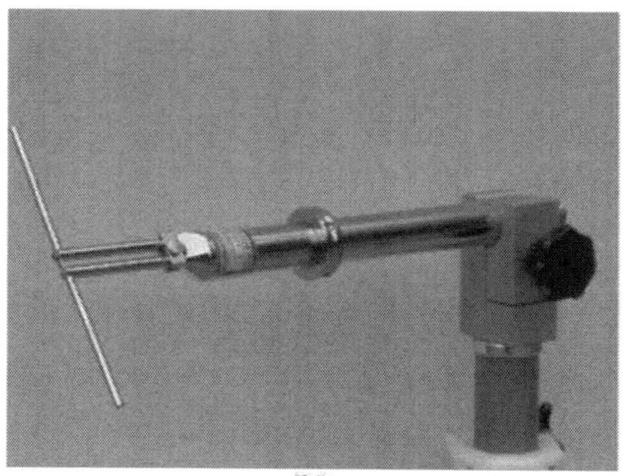

(b)

Figure 2. Antennae: (a) Monopole antenna derived from a coaxial cable by removing a part of outer conductor, (b) a UHF–Half–Wave Dipole (Wikipedia, 2011).

To accommodate the simulation of the dispersive soil and stability of the FDTD code at this frequency, the time increment Δt = 2 psec was used. The dispersive properties of the soil for this choice of center frequency and time-step are modeled with the Z-transform function coefficient set ε_{Ave}, a_1, b_0, b_1, and b_2, given in Table 2(b).

To relate the results to the field site, the model can be scaled up in size while scaling down the frequency. To evaluate the feasibility of the DNAPL detection method using monopole antennae, wave propagation through the background soil and scattered EM wave propagation by a DNAPL pool were modeled and analyzed. The geometry details of the monopole transmitting antenna modeled in this case are tabulated in Table 2(c). The drive signal excites the top of the simulated coaxial cable feeding the monopole antenna in a conventional radial field pattern. The electric field components on all grid points of different cross-sectional and depth slices of the medium were computed and then visualized using MATLAB.

First, the background medium was analyzed. Then, a rectangular acrylic plate as a representative of a DNAPL pool was modeled within the soil medium. Fig. 3 schematically shows the simulated geometry (the monopole antenna, the DNAPL pool, and the soil medium). Details of the geometry of the DNAPL pool scatterer are listed in Table 2(d).

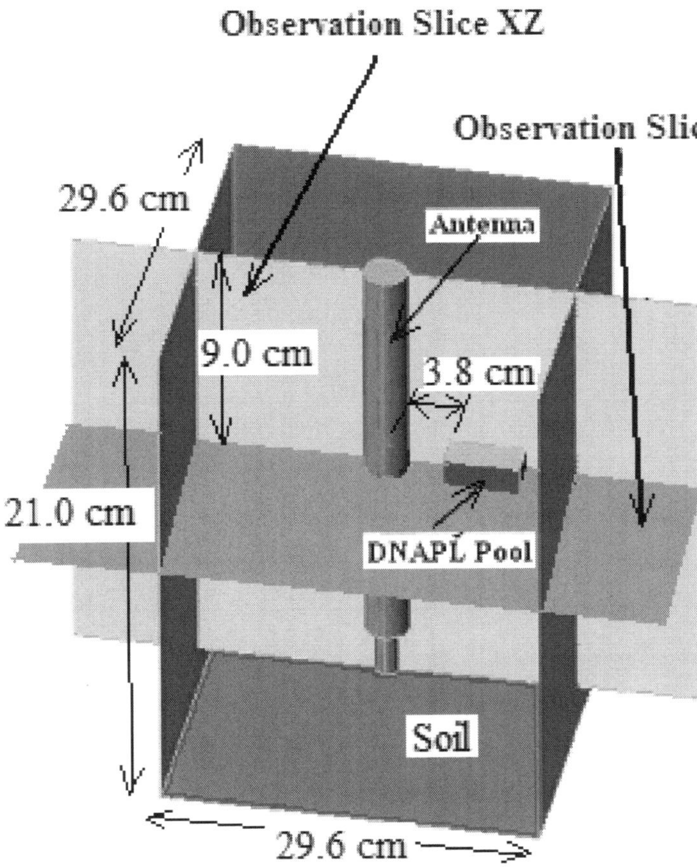

Figure 3. SoilBED, antenna, observation slice and rectangular DNAPL pool (3 cm × 3 cm × 1 cm).

Table 2a. Details of the simulated medium

Geometry	Size
Simulated grid	149 × 149 × 29
Grid cell size	0.2 cm × 0.2 cm × 1 cm
Entire grid size	29.6 cm × 29.6 cm × 28 cm
Soil thickness	21 cm
Air thickness	7 cm

Table 2b. Soil properties, used for the simulation of the fully saturated sandy soil at f = 1.5 GHz, t = 2 psec, and 17% gravimetric moisture content (w)

Parameter	Value
ε_{Ave} (Dielectric Permittivity)	20.9
a_1	-0.8985
b_0	-34.3627
b_1	68.7577
b_2	-34.3945

Due to solving the problem at Δt = 2 psec, the FDTD code is very sensitive, and all digits are necessary to satisfy the stability conditions

With respect to the center of the grid

To evaluate the wave propagation, the following observation slices were selected. Different components of electric field were computed and visualized on these slices.

Table 2c. Geometry details of the simulated monopole antenna

Antenna Details	Size
Antenna depth	120 mm
Perfectly conducting core wire thickness	1 mm
Extended dielectric length	20 mm
Extended dielectric thickness	3 mm
Perfectly conducting outer conductor (shield) thickness	3 mm
Frequency	1.5 GHz
Gaussian width	0.667 nsec
Gaussian peak	5 nsec

* The dielectric constant and effective electrical conductivity of the extended dielectric of the antenna are respectively assumed to be 2.1 and zero (Ω^{-1}).

Table 2d. Details of the DNAPL pool scatterer

DNAPL Pool Geometry	Size
Horizontal cross-section	3 cm × 3 cm × 1 cm
Depth	9 cm
Clear separation from the antenna	3.8 cm
Coordinate of the pool center*	6 cm, 0 cm, -2 cm

- A cross-sectional (horizontal: XY-plane) slice, cutting through the antenna and DNAPL pool at the depth of 9 cm. Z and X components of the electric field (E_z and E_x) are shown on this slice in Fig. 4.

- Up to this point, only the three vector components of the electric field were visualized. Now, the power is depicted. The intensity of a rapidly varying field is often displayed on a dB scale, enabling the visualization of small amplitude levels. This scale is given by 20 $\log_{10}|E / E_{max}|$. It is important to note that on the selected depth slice, E_y equals 0, and hence $\mathbf{E} = E_x \vec{i} + E_z \vec{k}$. In addition, since the time domain signals are all purely real, but may have positive or negative values, the dB scale is artificially augmented with positive values to indicate negative field values and better display the oscillating nature of the rapidly decaying wave. The sign of corresponding E_z governs the sign of the dB value. It should be stressed that 0 dB is the maximum field intensity, and positive dB values correspond to weaker signals with the opposite sign.

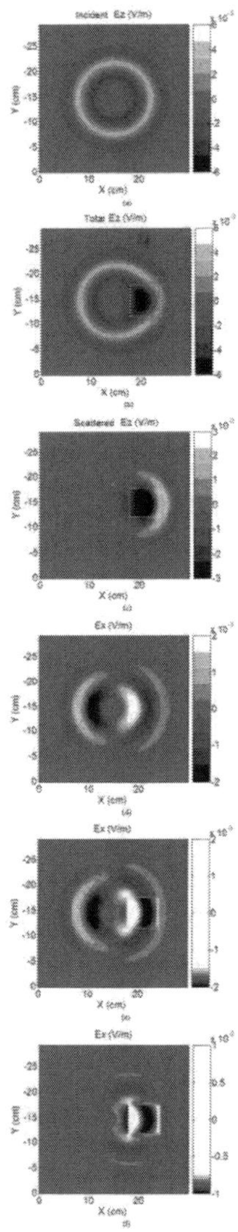

Figure 4. Electric field simulated on the cross-sectional slice (XY-plane) at t = 3.6 nsec (the extent of the DNAPL pool is marked by a yellow box): Z-component of the electric field: a) Incident, b) Total, and c) Scattered; and X-component of the electric field: d) Incident, e) Total, and f) Scattered.

A depth (vertical: XZ-plane) slice, passing through the antenna and DNAPL pool. This slice (XZ-plane) was chosen because the YZ-plane does not intersect the DNAPL pool. Due to symmetry, E_y is zero on this XZ slice, and hence |E| can be computed by only E_x and E_z. Results are shown in Fig. 5.

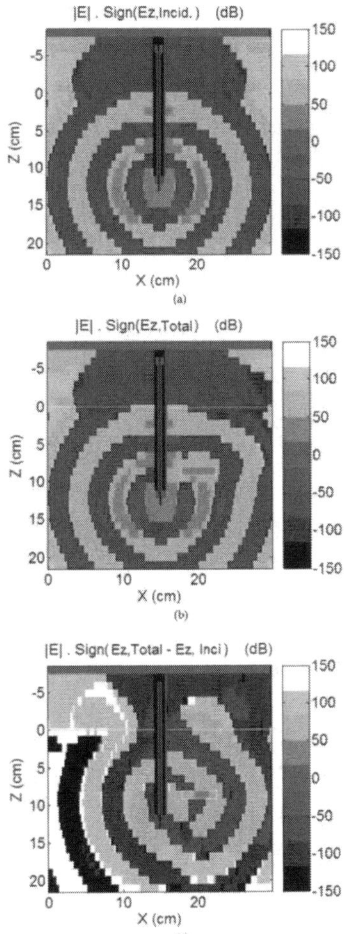

Figure 5. Electric field $[\mathrm{Sign}(E_{z,Total} - E_{z,Incident})] \times 20 \log_{10}(|E| \text{ or } |E_x \vec{i} + E_z \vec{k}|)$, on the depth slice (XZ-plane) at time t = 3.6 nsec (the extent of the DNAPL pool and soil-air interface are marked in yellow): a) Incident image, b) Total image, and c) Scattered image.

This case was initially analyzed without DNAPL contamination (incident field or background) and then with the DNAPL pool (total field). The scattered field by the DNAPL pool target can be computed by subtracting the two previous fields. Three figures are shown for each slice and for each electric field component and for: (i) "incident" (i.e., background, no target), (ii) "total" = background + DNAPL pool target as the scatterer, and (iii) "scattered" (i.e., signature of the target). All results shown in Fig. 4 are captured at t = 3.6 nsec. As seen, incident results of Figs. 4(a) and 4(d) are symmetric, while the total field results shown in Figs. 4(b) and 5(e) are not symmetric. The resulting scattered field information shown in Figs. 4(c) and 4(f) is asymmetric as well.

The incident, total, and scattered (target signature) fields are shown in Fig. 5. The monopole antenna was modeled as a Z-polarized antenna. Therefore, the Z-component of the electric field is the major component, but the scattered field by the DNAPL pool is also readily visible on the X and Y component plots. Since E_z dominates and the scattered field is visible on the Z-component (Fig. 5(c)), the scattered field shown on the dB plot will be clear as well. Further studies (that do not fit in this chapter) show weaker scattered Z-component in dry sandy soils. Different components can be experimentally measured using a receiving antenna with a different polarization (e.g., an X or Y polarized antenna, which is simply a monopole placed horizontally) than the Z-polarized (vertical) transmitting antenna. The scattered field is comparable to the incident field in this case. This potential can also be demonstrated in a different form as shown in Fig. 6.

This figure shows that there is a considerable magnitude and travel time difference between the total and incident fields received at a receiver located right above the DNAPL pool. The strong magnitude difference (more than 100%) and time difference (around 100 psec) between the two signals illustrate the potential of the cross-borehole GPR method to detect DNAPL pools. The early arrival of the total field is caused by the increase in the velocity of EM waves through the DNAPL pool due to its lower dielectric permittivity compared to the saturated soil. The increase in the magnitude of the total field is, on the other hand, caused by lower loss through the DNAPL pool due to its lower electrical conductivity. This illustrates a great potential for DNAPL detection using CWR in saturated soils, if the thickness and size of the pool is a reasonable fraction of the wavelength.

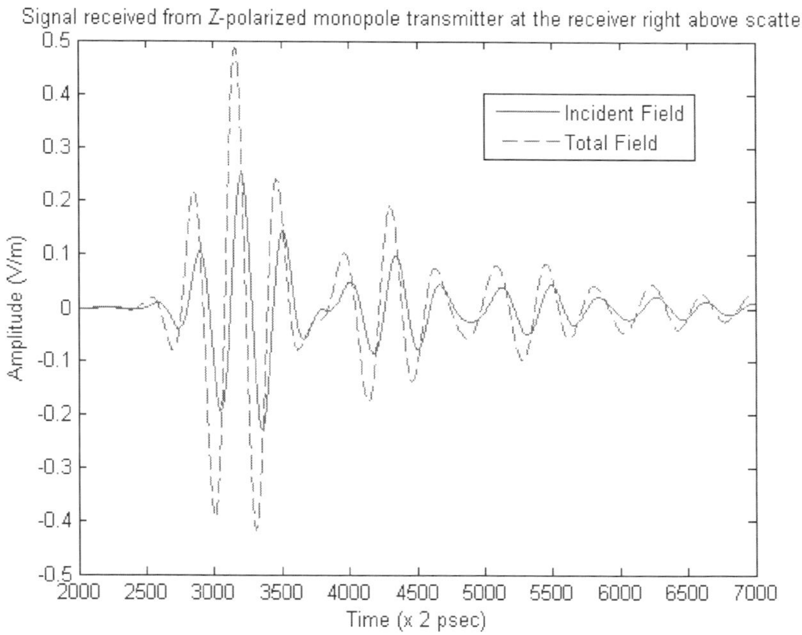

Figure 6. Z-component of total and incident electric fields due to the monopole antenna, received at a receiver located right above the DNAPL pool.

5.2. Dipole antenna case (B)

The above-mentioned small size monopole case can be scaled up to a more realistic size contaminated site. However, scaling up the results may cause some problems that do not allow a simple and direct generalization from small numerical models to real size contaminated sites. For example, in a non-dispersive medium, linear enlargement of the size can be simply interpreted to a linear increase in the wavelength and decrease in the frequency. However, in a dispersive medium, any change in the frequency causes variations in the dielectric properties of the medium. This change in the dielectric constant causes variations in the wave velocity, which in turn adds nonlinearity to the scaling process from the simulated medium up to the real size.

Therefore, to evaluate the scaling issues in a dispersive medium and study the effect of different radiation patterns of different antennae, another case with a more realistic size of soil medium surrounding a dipole antenna was modeled. The dipole is also larger than the monopole, since the smallest object to be modeled (the antenna) controls the uniform grid size in X and Y directions and size limitations of the FDTD code. The details about the grid size and the geometry of the soil medium for this case are tabulated in Table 3(a).

To decrease the computation cost, a much larger grid cell (3 cm in X and Y directions, and 5 cm in Z direction) was modeled (Table 3(a)). To satisfy sampling limitations (grid size < λ / 10) and study the scaling effect, the wavelength should be larger. Therefore, the frequency was selected to be 100 MHz (lower than 1.5 GHz in Case A). To satisfy the Courant's condition for the new grid size, the time increment was increased to Δt = 50 psec.

Table 3a. Details of the simulated medium

Geometry	Size
Simulated grid	149 × 149 × 69
Grid cell size	3 cm × 3 cm × 5 cm
Entire grid size	444 cm × 444 cm × 340 cm
Soil thickness	305 cm
Air thickness	35 cm

The soil medium is exactly the same fully water-saturated sandy soil modeled in the previous case with 17% gravimetric moisture content. However, dielectric properties of the dispersive soil at the different frequency and time increments (f = 100 MHz, and Δt = 50 psec) are different. Therefore, the dielectric constant and coefficients (ε_{Ave}, a_1, b_0, b_1, and b_2) of the Z-transform function required to model the dispersive electrical conductivity of the soil were recomputed for the new frequency and time increment. The new soil parameters are listed in Table 3(b). A center-fed resistively tapered ½ wavelength dipole antenna is modeled as the transmitter. The particular details of the resistive dipole are avoided by modeling the antenna electromagnetically as simply a tapered half-wave surface current source residing on the exposed coaxial insulator (maximum at the center, the point where the feed line joins the elements, and zero at the ends of the elements). This type of antenna may be used in a PVC-lined borehole filled with water. Therefore, the model simulates the antenna surrounded by water. Obviously, to model the dispersive nature of water and maintain the symmetry and accuracy on the circular interface around the antenna, water is modeled using the same technique used to model lossy dispersive soils (Weedon & Rappaport, 1997). For the same reason, the dielectric portion is modeled using the same technique used for lossy dispersive soils, despite the non-lossy and non-dispersive nature of the dielectric material. The PVC casing was ignored during the simulation to simplify the modeling and because the wall of the PVC is very thin compared to the wavelength (780 mm) of the EM wave. The dipole antenna is Z-polarized and the excitation signal is a 100 MHz cosine-modulated Gaussian pulse, progressively delayed along the antenna in the Z-direction (i.e., points along the Z-directed dipole are excited with a progressive phase delay proportionate to the traveling time of the current fed through the midpoint and along the dielectric portion of the dipole). Table 3(c) summarizes the details about the structure of the dipole antenna.

Table 3b. Soil parameters, used for the simulation of the fully saturated sandy soil at f = 100 MHz, Δt = 50 psec, and 17% gravimetric moisture content

Parameter	Value
ε_{Ave} (Dielectric Permittivity)	14.9251
a_1	-0.8985
b_0	1.04948
b_1	-1.9896
b_2	0.94093

Due to solving the problem at Δt = 50 psec, the FDTD code is very sensitive, and all digits are necessary to satisfy the stability conditions.

Table 3c. Details of the simulated dipole antenna

Antenna Details	Size
Antenna depth	1800 mm
Borehole diameter	240 mm
Perfectly conducting core wire thickness	22 mm
Extended dielectric length	500 mm
Extended dielectric thickness	64 mm
Perfectly conducting outer conductor (shield) thickness	43 mm
Depth of water-filled borehole	500 mm
Frequency	100 MHz
Gaussian width	10 nsec
Gaussian peak	75 nsec

* The dielectric constant and effective electrical conductivity of the extended dielectric are respectively assumed to be 2.1 and zero (Ω-1).

First, the wave propagation through the soil background was analyzed. Then, a rectangular DNAPL pool was modeled within the soil medium. Fig. 7 schematically shows the geometry of this DNAPL pool. The details about the geometry of the DNAPL pool scatterer are listed in Table 3(d).

Table 3d. Details about the DNAPL pool scatterer

DNAPL Pool Geometry	Size
Horizontal area	45 cm × 45 cm × 15 cm
Depth	90 cm
Clear distance to the antenna	22.5 cm
Coordinate of the pool center*	90 cm, 0 cm, 45 cm

* With respect to the center of the grid

Similar to Case A, the transmitting antenna is modeled in the code, but rather than modeling receiving antennae, the three different components of the electric and magnetic fields are computed at all grid points on the following depth and cross-sectional slices.

- A cross-sectional (horizontal: XY-plane) slice, cutting through the antenna and DNAPL pool at the depth of 90 cm. Z and X components of the electric field (E_z and E_x) are shown on this slice (Fig. 8).

- A depth (vertical: XZ-plane) slice, passing through the antenna and DNAPL pool. The magnitude of the power, derived from both E_x and E_z, is shown on this slice in Fig. 9 (E_y is zero on this slice due to symmetry).

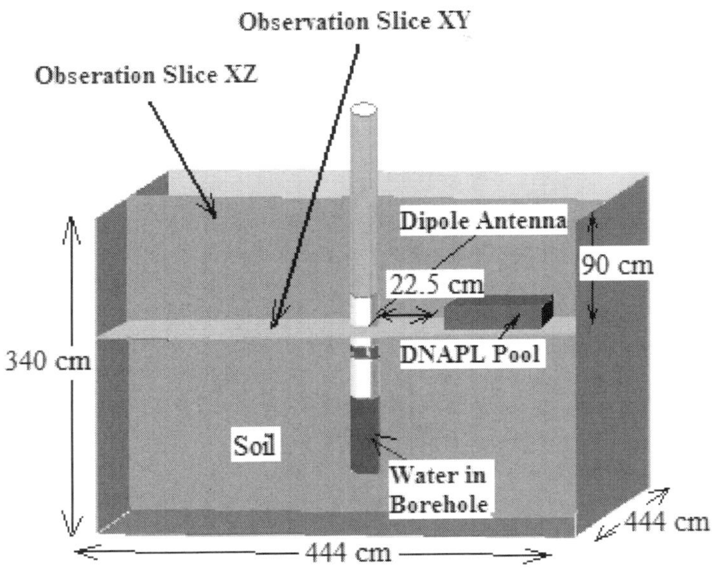

Figure 7. Schematic representation of the borehole dipole antenna geometry and DNAPL pool (45 × 45 cm × 15 cm).

As before, for each slice and each electric field component, three figures are shown: (i) incident (background) field, (ii) total field, and (iii) scattered field. As seen in Figs. 8(a) and 8(d), background results are symmetric. The total fields of Figs. 8(b) and 8(e) and the scattered field shown in Figs. 8(c) and 8(f) are asymmetric. The interesting and encouraging point is the visibility of the DNAPL pool over the entire medium within the total and scattered fields, even on the far side of the pool from the antenna. This predicts that dipole antenna boreholes of the CWR method can be drilled far from DNAPL-contaminated zones to reduce the risk of further vertically downward penetration of DNAPLs associated with drilling through contaminated zones. This appears to be more valid at higher degrees of water-saturation. Again, this potential can also be demonstrated in a different form as shown in Fig. 9.

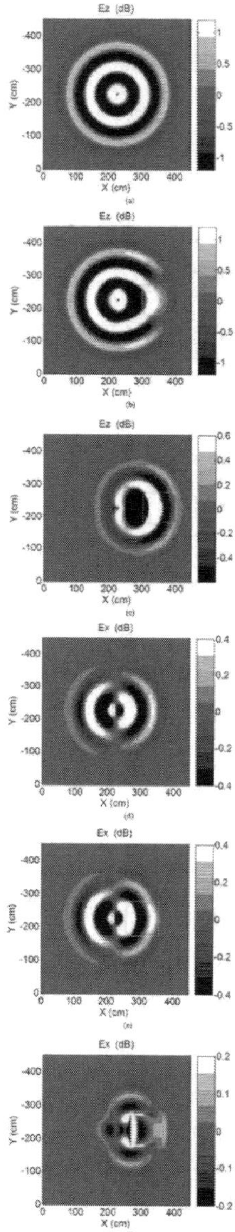

Figure 8. Electric field simulated on the cross-sectional slice (XY-plane) at t = 90 nsec (the extent of the DNAPL pool is marked by a yellow box): Z-component of the electric field: a) Incident, b) Total, and c) Scattered X-component of the electric field: d) Incident, e) Total, and f) Scattered.

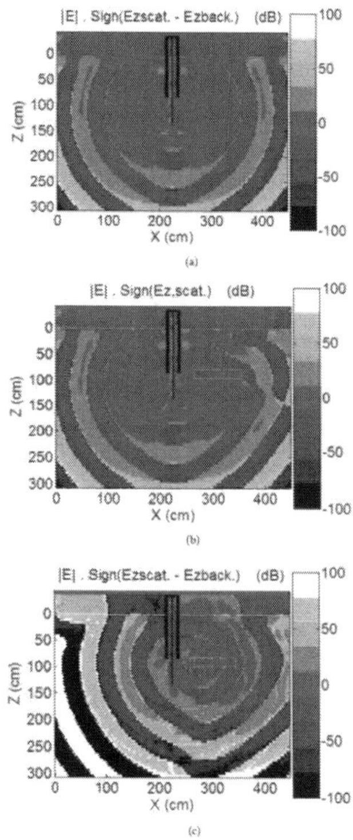

Figure 9. Electric field, $[Sign(E_{z,Total} - E_{z,Incident})] \times 20 \log_{10}(|E| \; or \; |E_x \vec{i} + E_z \vec{k}|)$, on the depth slice (XZ-plane) at time t = 90 nsec (the extent of the DNAPL pool and soil-air interface are marked in yellow): a) Incident image, b) Total image, and c) Scattered image.

Previously, in the case of the monopole transmitter, the received total and incident signals were computed at a receiver located right above the DNAPL pool. Now, the two are computed for a receiver located 175 cm above the pool to examine the possibility of minimizing the destructive effect of placing the receiving antenna too close to the pool. This figure again demonstrates a strong magnitude and travel time difference between the total and incident fields received at the receiver located far above the DNAPL pool. The strong magnitude difference (around 40%) and time difference (around 2.5 nsec.) between the two signals, once again, embraces the potential of the use of the cross-borehole GPR method to detect DNAPL pools. As in the monopole case, the early arrival of the total field and its higher magnitude can be respectively justified by the higher velocity of EM waves through the DNAPL pool due to its lower dielectric permittivity (relative to the water-saturated soil) and the lower loss due to the pool's lower electrical conductivity. The dry soil case has been studied (it does not fit in the extent of this chapter) and proved to have weaker scattered signals, but still strong enough to have the potential to detect DNAPL pools using relatively widely spaced antennae.

As expected for the radiation pattern of dipoles, most of the energy is transmitted perpendicularly to the antenna through the mid-part of the antenna into the soil. The modeled dipole antenna is Z-polarized, and thus the Z-component of the electric field is the major component. The monopole antenna behaves similarly to the dipole antenna, the only difference being the strong signature of the DNAPL pool on the Z-component as well as X and Y components in the case of the dipole antennae. This wider spread perturbation due to the clutter promises more potential detection key-points for the dipole antenna installation. It is noteworthy that the perturbation due to the object on the X and Y components of the electric field of both monopole and dipole antennae appears more to the sides of the contaminated zone and perpendicular to the line connecting the center of the contaminated zone and the antennae. The perturbation on the Ez component in the monopole antenna case is distributed throughout the contaminated zone, while for the dipole one, this perturbation spreads to the far side across the contaminated zone as well.

6. COMPARISON WITH MONOPOLE EXPERIMENTATION

In this section, the numerically simulated monopole case is compared with experimentation for validation. The experimental setup (for more information refer to Farid et al., 2006) uses two PVC-cased ferrite-bead-jacketed monopole antennae connected to a vector network analyzer (Agilent 8714ES), and frequency-response measurements were collected for a homogeneous water-saturated sandy soil background. Fig. 11 shows a schematic of the experiment.

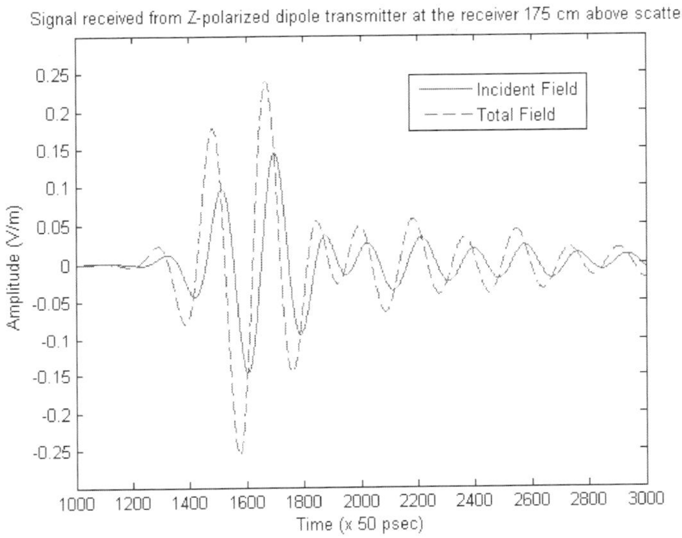

Figure 10. Z-component of total and incident electric fields due to the dipole antenna, received at a receiver located 175 cm above the DNAPL pool.

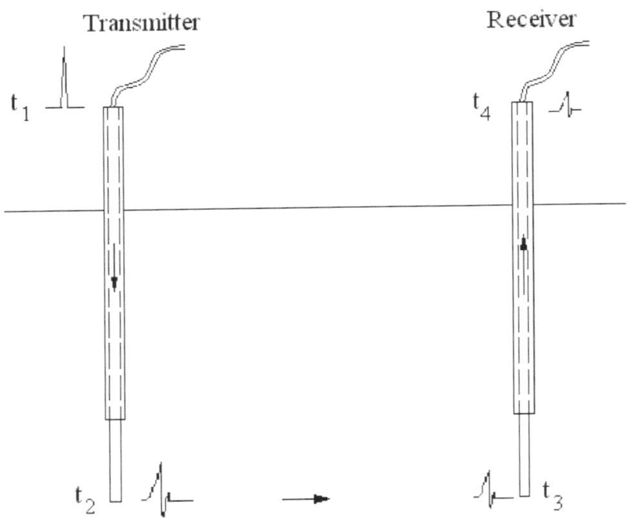

Figure 11. Pulse traveling through transmitter, soil and then receiver

The frequency-response measurements are collected across the frequency range of 0.4 to 2.2 GHz. The frequency-response measurements are, then, transformed to the time domain using an inverse fast Fourier transform (IFFT) and an assumption of a narrow-width, wideband Gaussian pulse as the transmitted signal. Both the experiment and the FDTD model use the same Gaussian pulse source. Due to the frequency range used in the experimentation (0.4 GHz to 2.2 GHz), the width of the Gaussian signal should not exceed a maximum of: $W_{min.} = 1 / f_{max.} = 1 / 2.2$ GHz = 0.455 nsec. Using narrower signals seems to create a noise. Therefore, the upper limit (0.455 nsec.) is selected as the width of the signal. Fig. 12 shows the transmitted signal.

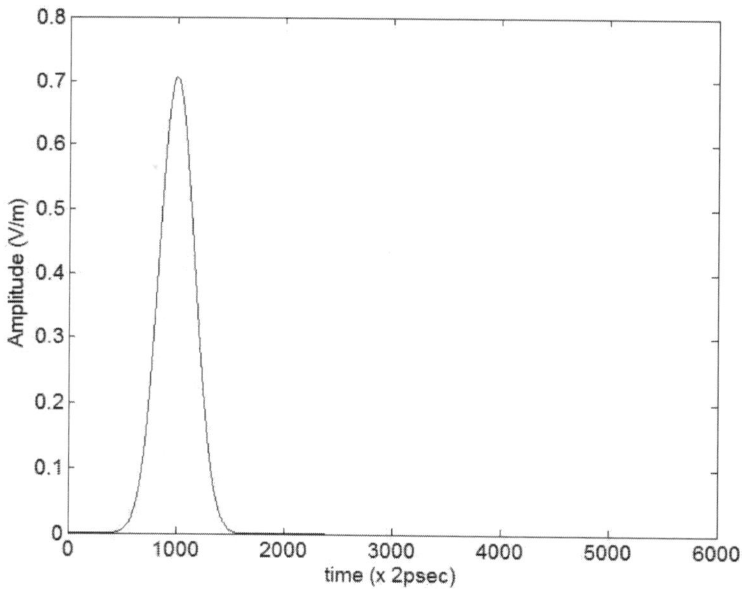

Figure 12. X- and Y-components of the radial excitation at the end of the feed cable into the top of the transmitting antenna ($E_{x1} = E_{y1}, = 0.7$ V/m, $E_{z1} = 0$, and $|E| = 1$ V/m)

The FDTD-simulated received signal at the bottom of the receiving antenna is shown in Fig. 13(a). As seen, this received signal is distorted and does not resemble the Gaussian transmitted one. Therefore, its peak is not easily distinguishable, since the received signal is modulated and noisy. To resolve this issue, the received signal should first be demodulated and then low-

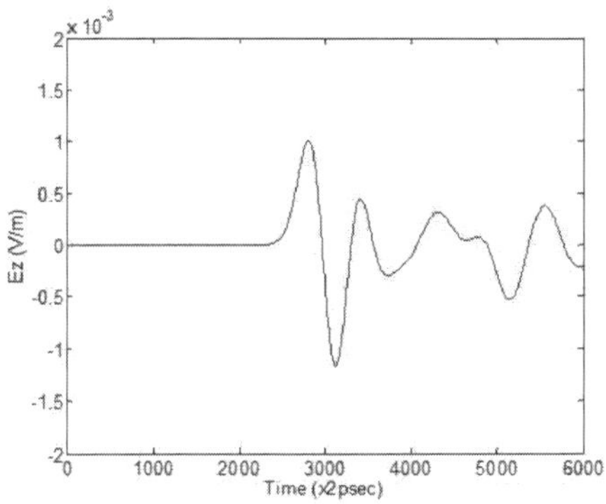

(a)

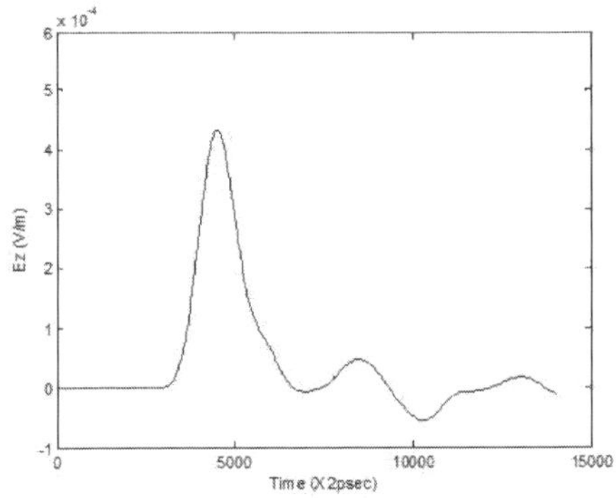

(b)

Figure 13. Received signal (E_{z3}) at the bottom of the receiver in the saturated background from the FDTD simulation at time t_3: a) Unprocessed, and b) Processed.

pass-filtered. The demodulation frequency can be found by observing the received signal in the frequency domain (computed via a fast Fourier transform). This demodulation frequency is observed to be dependent on the separation between the transmitting and receiving antennae. A MATLAB code was prepared to automatically observe the received signals in the frequency domain, find the proper demodulation frequencies, and find the proper low-pass filter to filter the noise. The processed received signal is shown in Fig. 13(b).

The travel time of the FDTD model can be simply calculated as the difference between the peak times of the received (Fig. 13(b)) and transmitted (Fig. 12) signals.

To transform the experimental frequency-response to the travel-time, the transmitted signal is fast Fourier transformed to the frequency domain and multiplied by the frequency-response (Fig. 14) to find the received signal at the receiver in the frequency domain. Then, the result is transformed back to the time domain using an inverse fast Fourier transform. The result (received signal in the time domain) is shown in Fig. 15(a).

This signal does not resemble the transmitted Gaussian signal. Therefore, it needs to be processed (demodulated and low-pass-filtered). The processed received signal is shown in Fig. 15(b). The demodulation frequency and filter design vary with the distance between the transmitting and receiving antennae, which is automatically calculated using the above-mentioned MATLAB code.

The travel time of the experimental model can be simply calculated as the difference between the peak times of the received (Fig. 15(b)) and transmitted (Fig. 12) signals.

Since the vector network analyzer is calibrated at the end of the cables (the connection points to the monopole antennae), the experimental travel-time is measured between these two points. On the other hand, the FDTD travel-time ($t_3 - t_1$) is computed between the feed cable and the bottom of the receiving antenna. Therefore, it needs to be adjusted for this difference.

Up to this point, the FDTD travel-time ($t_3 - t_1$) from the feed cable to the tip of the receiving antenna is computed. The travel time through the receiving antenna ($t_4 - t_3$), which is by symmetry equal to ($t_2 - t_1$), should be added to ($t_3 - t_1$) to find the total travel time between the feed and receiver cables ($t_4 - t_1$) for the FDTD model. The resulting travel time from the FDTD simulation can be used for comparison with the experimental results.

The travel time computed from the forward model is (4500 + 900 - 1000) × 2 psec = 8.8 nsec, which closely agrees with the one indirectly computed from the experimentally collected frequency-response data: (5700 - 1000) × 1.87 psec = 8.6 nsec. The difference is due to the slight, potential discrepancy between the dielectric constant assigned to the forward model (used from the results of another work by the authors (Zhan et al., 2007)) and the real values of the experimentation.

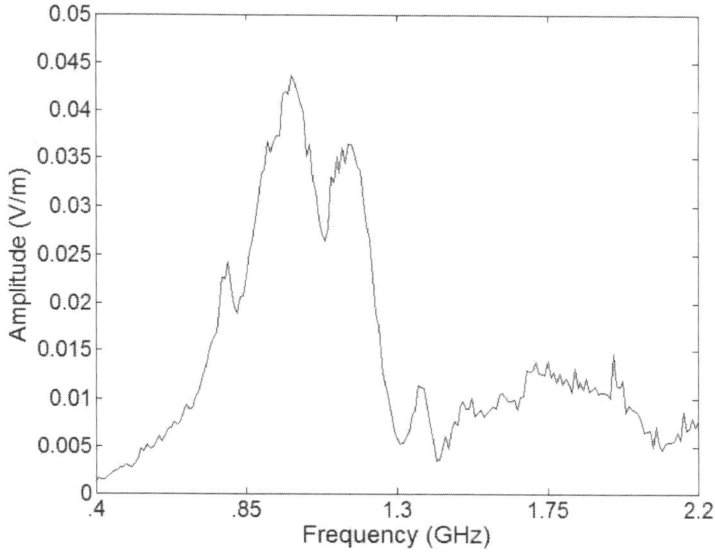

Figure 14. Experimental frequency-response in water-saturated background soil

The intensities of the unprocessed received signals from the FDTD simulation (Fig. 13(a)) and experimentation (Fig. 15(a)) agree relatively well, but not perfectly. The reason is the potential slight discrepancy between the electrical conductivity assigned to the FDTD model compared to the actual one of the experiment. However, due to the difference between the necessary processing methods (different filters), the intensity of the processed received signals for the FDTD simulation (Fig. 13(b)) and the one of the experiment (Fig. 15(b)) do not agree as closely.

This comparison consists of the incident field for the homogeneous background soil. The comparison for the total and scattered fields at the presence of any anomalies (e.g., dielectric objects) will be conducted in the future.

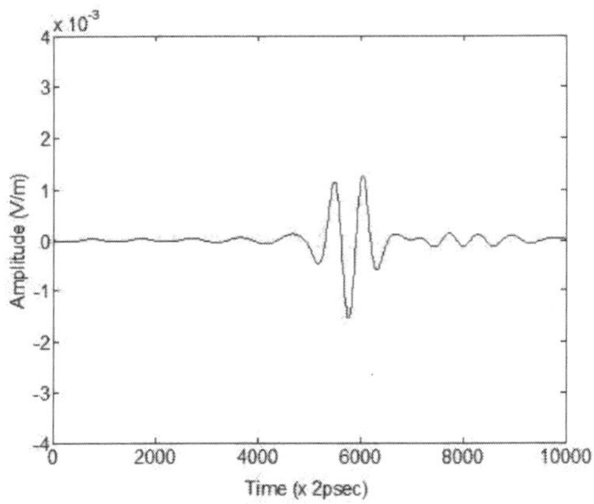

(a)

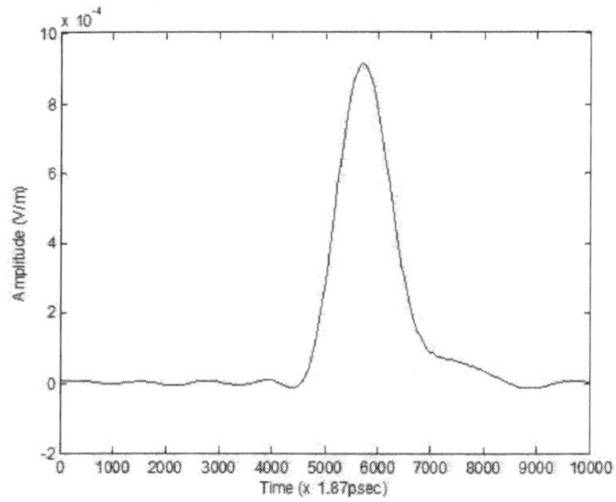

(b)

Figure 15. Received signal (E_{z4}) at the top of the receiver in the saturated background, indirectly computed from the experimental frequency-response: a) Unprocessed, and b) Processed.

7. CONCLUSION

A finite difference time domain (FDTD) model was developed for monopole and dipole antennae. Then, the scattering due to dielectric materials (to simulate DNAPL pools) in soils was modeled and analyzed. Results of the two simulated cases using the FDTD model demonstrate strong perturbation by the DNAPL pool on the electric field in the fully water-saturated sandy soil. In the case of the monopole antenna, the DNAPL pool target is more visible on the X and Y components of the electric field compared to the major component Z. The perturbation on the intensity of the electric field (|E|) transmitted by the monopole antenna is not as strongly visible as in the dipole case. In the dipole case, X and Y components are those parallel to likely hydraulic-conductivity contrast planes (e.g., usually horizontal clay lenses within a thick sand layer), which are potential locations to accumulate DNAPLs.

Different components of the electric field can selectively be collected using receiving antennae with different polarizations from the polarization of the transmitting antenna (e.g., a horizontally-polarized receiving monopole antenna and a vertically-polarized transmitting monopole antenna). Therefore, designing the receiving antenna alignment and polarization to selectively collect electric field components parallel to a possible DNAPL pool may help to compensate for a stronger perturbation on the minor components (X and Y) of the electric field emitted from a Z-polarized monopole antenna. These minor components should be of a high enough signal to noise ratio.

In the case of the dipole antenna, all three components of the electric field in the fully water-saturated soil have almost equal detection potential. In both of the above cases, there is a strong dielectric contrast between the DNAPL pool and the water-saturated soil. However, different radiation patterns of the dipole antenna compared to the monopole antenna may make the dipole antenna more desirable for DNAPL detection.

Field problems can be scaled down in size along with scaling up the frequency in non-dispersive soils to achieve the proper geometry and frequency for simulation purposes. This linear scaling of frequency and size may not work as well for dispersive soils, since frequency-dependent dielectric properties of dispersive soils add nonlinearity to the scaling problem. Other conclusions follow.

- Images provided by such simulations show the field distribution that exists throughout the subsurface (i.e., similar to filling the entire volume with receiver antennae), but the field can only be observed practically by placing a reasonable number of receiving antennae at key underground positions with the appropriate polarization. This research can be used to find the radiation patterns of different antenna types and the interaction of the radiated field with soil heterogeneities, which leads to a better understanding of subsurface wave behavior at these key positions and aids the selection of optimum antenna patterns to cover these key positions.

- While the depth of contamination is a problem for surface-reflection methods (e.g., GPR), there are no theoretical depth limitations for CWR, except practical drilling limitations and cost. The separation limitations between transmitting and receiving antennae used for CWR still exist. However, CWR has the advantage of using a one-way traveling path (transmission), unlike the two-way traveling path of surface-reflection GPR. In addition, the strong reflecting air-soil interface in the surface-reflection GPR technique is eliminated in the CWR technique and replaced with a better-controlled coupling between the borehole antennae and surrounding soil.

- The perturbation due to the DNAPL target is stronger for the greater dielectric permittivity contrast between DNAPL pools and highly moist soil, as opposed to DNAPL plumes with low DNAPL saturation and dryer soils.

- The signal to noise ratio of the scattered field by DNAPL pools should be high enough for measurements. As seen in the figures, the scattered field is comparable to the incident field. Therefore, if the signal to noise ratio of the incident field is high enough for measurement, the scattered field will probably have a large enough signal to noise ratio to be measurable as well.

- The results of this forward model with monopole and dipole antennae show that the field perturbation (scattered = total - incident) for relatively large DNAPL pools at high enough DNAPL saturation, is of the same order of magnitude as the incident signal. This proves DNAPL detection using CWR in water-saturated soils feasible. The simulation tool can also be used as a forward model to develop an inverse scheme for DNAPL imaging.

- Armed with the background data as well as the radiation patterns of different antennae (via simulations like those in this chapter), the existence of DNAPL pools can be confirmed with efficient inverse models and judicious placement of receiving antennae (i.e., pattern of antenna installation) where stronger perturbation and reception by receiving antennae are expected.

CWR may be a feasible and reasonable method to monitor DNAPL pools in a suitable environment. This most suitable environment is a medium consisting of a low-loss, low-heterogeneity porous material. In other media, it is more difficult to distinguish DNAPL accumulation from geologic variations, which are more complicated due to heterogeneity. Nevertheless, soil heterogeneity may not pose a crucial problem under water-saturated conditions since different soils behave similarly at relatively high degrees of water-saturation and high frequencies (the case is different for low frequencies). Monitoring DNAPL movement may well be possible or easier in an even less saturated heterogeneous environment because of the static nature of stratigraphic events and the dynamic nature of DNAPL flow. Several features of DNAPL pools may help to distinguish them from stratigraphic events, such as their irregular shapes with sharp lateral boundaries.

Finally, the FDTD model was compared for the incident field due to the monopole case in a homogeneous water-saturated sandy soil background with the experimental results. The reasonable agreement between both the travel time and intensity of the unprocessed, simulated and experimental results validates the FDTD model. The comparison and validation for the total and scattered fields at the presence of any anomalies (e.g., dielectric objects) need to be studied in the future.

8. ACKNOWLEDGEMENTS

This research was supported in part by the Gordon Center for Subsurface Sensing and Imaging Systems (CenSSIS), under the Engineering Research Centers Program of the National Science Foundation (NSF: Award Number EEC-9986821).

The authors would like to express gratitude for financial and scientific support provided by the Gordon CenSSIS and NSF.

REFERENCES

1. Ajo-Franklin, J. B., Geller, J. T. & Harris, J. M. (2004). The dielectric properties of granular media saturated with DNAPL/water

mixtures. *Geophysical Research Letters (GRL)*, Vol. 31, No. 17, L17501

2. Anderson, J. & Peltola, J. (1996). *Ground Penetrating Radar as a tool for detecting contaminated areas: Groundwater Pollution Premier*, CE 4594 Soil and Groundwater Pollution, Civil Engineering Department, Virginia Tech., Date of Access: Feb/2011, Available from: <http://www.cee.vt.edu/program_areas/environmental/teach/gwprimer/gprjp /gprjp.html#Intro>

3. Arulanandan, K. (1964). Dielectric method for prediction of porosity of saturated soils. *ASCE Journal of Geotechnical Engineering*, Vol. 117, No. 2, pp. 319-330

4. Arulanandan, K. & Smith, S. S. (1973). Electrical dispersion in relation to soil structure. *ASCE Journal of Soil Mechanics and Foundation Div.*, Vol. 99, No. 12, pp. 1113-1133

5. Balanis, C. A. (1989). *Advanced engineering electromagnetic,* John Wiley & Sons, New York, 1008p

6. Belli, K., Rappaport, C., Zhan, H. & Wadia-Fascetti, S. (2009). Effectiveness of 2D and 2.5D FDTD Ground Penetrating Radar Modelling for Bridge Deck Deterioration Evaluated by 3D FDTD. *IEEE Transactions on Geoscience and Remote Sensing*, Vol. 47, No. 11, pp. 3656 – 3663

7. Belli, K., Rappaport, C. & Wadia-Fascetti, S. (2009a). Forward Time Domain Ground Penetrating Radar Modelling of Scattering from Anomalies in the Presence of Steel Reinforcements. *Research in Nondestructive Evaluation*, Vol. 20, No. 4, pp. 193 - 214.

8. Binley, A., Winship, P. & Middleton, R. (2001). High resolution characterization of vadose zone dynamics using Cross-Borehole Radar. *Water Resource Research*, Vol. 37, no. 11, pp. 2639-2652

9. Blackhawk Geoservices Inc. (2008). *Integrated geophysical detection of DNAPL source zones,* Final Report, Date of Access: Feb/2011, Available from: http://handle.dtic.mil/100.2/ADA409159

10. Bradford, J. H. & Wu, Y. (2007). Instantaneous spectral analysis; time-frequency mappingvia wavelet matching with application to

3D GPR contaminated site characterization. *The Leading Edge*, Vol. 26, pp. 1018-1023

11. Brewster, M. L. & Annan, A. P. (1994). GPR monitoring of a controlled DNAPL release, 200 MHz radar. *Geophysics*, Vol. 59, No. 8, pp. 1211-1221

12. Daniels, J. J., Roberts, R. & Vendl, M. (1992). Site studies of Ground Penetrating Radar for monitoring petroleum product contaminants. *Proceedings of SAGEEP (Symposium of the Applications of Geophysics to Engineering and Environmental Problems)*, Society of Engineering Mine Exploration, pp. 597–609

13. Dobson, M. C., Ulaby, F. T., Hallikainen, M. T. & El-Rayes, M. A. (1985). Microwave dielectric behavior of wet soil, Part II: Dielectric mixing models. *IEEE Transaction on Geoscience and Remote Sensing*, GE- 23, No. 1, pp. 35–46

14. Farid, A. M., Alshawabkeh, A. N. & Rappaport, C. M. (2006). Calibration and Validation of a Laboratory Experimental Setup for CWR in Sand. *ASTM, Geotechnical Testing Journal*, Vol. 29, Issue 2, pp. 158-167

15. Firoozabadi, R., Miller, E., Rappaport, C. & Morgenthaler, A. (2007). A New Inverse Method for Subsurface Sensing of Object under Randomly Rough Ground Using Scattered Electromagnetic Field Data. *IEEE Transactions on Geoscience and Remote Sensing*, Vol. 45, No. 1, pp. 104-117.

16. Gandhi, O. (1993). A frequency dependent FDTD (Finite Difference Time Domain) formulation for general dispersive media. *IEEE Transactions on Microwave Theory and Techniques*, Vol. 41, pp. 658-665

17. Geller, J. T., Kowalsky, M. B., Seifert, P. K. & Nihei, K. T. (2000). Acoustic detection of immiscible liquids in sand. *Geophysical Research Letters*, Vol. 27, No. 3, pp. 417-420, 2000

18. Grant, I. S. & Philips, W. R. (1990). *Electromagnetism,* John Wiley & Sons, New York, 525 pp

19. Grimm, R. & Olhoeft, G. (2004). Cross-hole complex resistivity survey for PCE at the SRS A-014 outfall. *Proceedings of SAGEEP*, Colorado Springs, Colorado, 2004,pp. 455-464

20. Hallikainen, M. T., Ulaby, F. T., Dobson, M. C., El-Rayes, M. A. & Lin-Kun, W. (1985). Microwave dielectric behavior of wet soil: Part I-Empirical models and experimental observations. *IEEE Transaction on Geoscience and Remote Sensing*, GE- 23, No. 1, pp. 25–34

21. Hipp, J. (1974). Soil electromagnetic parameters as functions of frequency, soil density, and soil moisture. *Proceedings of IEEE*, Vol. 62, No.1, pp. 98-103

22. Hoekstra, P. & Doyle, W. T. (1971). Dielectric relaxation of surface adsorbed water. *Journal of Colloid and Interface Science*, Vol. 36, No. 4, pp. 513-521

23. Hoekstra, P. & Delaney, A. (1974). Dielectric properties of soils at UHF and microwave frequencies. *Journal of Physics Research*, Vol. 79, No. 11, pp. 1699–1708

24. Interstate Technology and Regulatory Cooperation (ITRC) (2000). *Work Group DNAPLs/Chemical Oxidation Work Team, Dense Non-Aqueous Phase Liquids (DNAPLs)*, Review of Emerging Characterization and Remediation Technologies, Technology Overview, http://www.itrcweb.org/DNAPL-1.pdf

25. Kashiwa, T. & Fukai, I. (1990). A treatment by the FDTD method for the dispersive characteristics associated with electronic polarization. *Microwave and Guided Wave Letters*, Vol. 16, pp. 203-205

26. Kosmas, P. (2002). *Three-dimensional finite difference time domain modeling for Ground Penetrating Radar applications,* M.Sc. Thesis, Northeastern University, Boston,MA

27. Kunz, K. & Luebbers, R. (1993). *The FDTD (Finite Difference Time Domain) method for electromagnetic,* CRC Press, Boca Raton, Florida

28. Mur, G. (1981). Absorbing boundary conditions for the finite-difference approximation of the time-domain electromagnetic field equations. *IEEE Transactions on Electromagnetic Compatibility*, EMC- 23, pp. 377-382

29. Rappaport, C. M. & Winton, S. (1997). Modeling dispersive soil for FDTD computation by fitting conductivity parameters. *12th Annual Review of Progress in Applied Computational Electromagnetics Symposium Digest*, pp. 112-118

30. Rappaport, C. M., Wu, S. & Winton, S. C. (1999). FDTD wave propagation in dispersive soil using a single pole conductivity model. *IEEE Transactions on Magnetics*, Vol. 35, pp. 1542-1545

31. Rinaldi, V. A. & Francisca, F. M. (1999). Impedance analysis of soil dielectric dispersion (1 MHz – 1 GHz). *ASCE Journal of Geotechnical and Geoenvironmental Engineering*, Vol. 125, No. 2, pp. 111-121

32. Sachs, S. B. & Spiegler, K. S. (1964). Radio-frequency measurements of porous plugs, ion exchange resin-solution systems. *Journal of Physical Chemistry*, Vol. 68, pp. 1214-1222

33. Selig, E. T. & Mansukhani, S. (1975). Relationship of soil mixture to the dielectric property. *ASCE Journal of Geotechnical Division*, Vol. 101, No. 8, pp. 755–770

34. Sen, P. N., Scala, C. & Cohen, M. H. (1981). A self-similar model for sedimentary rocks with application to the dielectric constant of fused glass beads. *Geophysics*, Vol. 46, pp. 781-795

35. Sheriff, R. E. (1989). *Geophysical methods,* Prentice Hall, New Jersey, 605 p

36. Smith-Rose, R. L. (1933). The electrical properties of soils for altering current at radio frequencies. *Proceedings of Royal Society*, Vol. 140, No. 841A, pp. 359-377

37. Smith-Rose, R. L. (1935). The electrical properties of soils at frequencies up to 100 megacycles per second; with a note on resistivity of ground in the United Kingdom. *Proceedings of Physical Society*, Vol. 47, No. 262, pp. 923-931

38. Sneddon, K. W., Olhoeft, G. R. & Powers, M. H. (2000). Determining and mapping DNAPL saturation values from noninvasive GPR measurements. *Proceedings of SAGEEP*, Arlington, Virginia, pp. 293-302

39. Talbot, J. & Rappaport, C. M. (2000). An efficient Mur-type ABC for lossy scattering media. *Progress in Electromagnetics Research Symposium*, Vol. 194

40. Thevanayagham, S. (1995). Frequency-domain analysis of electrical dispersion of soils. *ASCE Journal of Geotechnical Engineering*, Vol. 121, No. 8, pp. 618-628

41. Von Hippel, A. R. (1953). *Dielectric materials and applications,* Technology Press of M.I.T. and John Wiley, New York

42. Weast, R. C. (1974). *CRC Handbook of Chemistry and Physics,* 55th edition, CRC Press, Cleveland, OH

43. Weedon, W. & Rappaport, C. M. (1997). A general method for FDTD modeling of wave propagation in arbitrary frequency-dispersive media. *IEEE Transactions on Antenna and Propagation*, pp. 401-410

44. Wikipedia, Date of Access: Feb/2011, Available from: http://en.wikipedia.org/wiki/Dipole_antenna

45. Yee, K. (1966). Numerical solution of initial boundary value problems involving Maxwell's equations in isotropic media. *IEEE Transaction on Antennae and Propagation*, Vol. 14, No. 3, pp. 302-307

46. Zhan, S. H., Farid, A., Alshawabkeh, A. N., Raemer, H. & Rappaport, C. M. (2007). Validated Half-Space Green's Function Formulation for Born Approximation for Cross-Well Radar Sensing of Contaminants. *IEEE, Transaction of Geoscience and Remote Sensing*, Vol. 45, No. 8, pp. 2423-2428, August

2

CHAPTER

ELECTROMAGNETIC WAVE PROPAGATION INTO FRESH WATER

Shan Jiang, Stavros Georgakopoulos

Department of Electrical and Computer Engineering, Florida International University, Miami, USA.

ABSTRACT

Wireless communications from air to fresh water are studied here. Our analysis relies on plane wave propagation models. Specifically, the transmission loss and propagation loss of RF waves penetrating into fresh water are calculated for various propagation depths. Even though RF wireless communications are not well suited for seawater due to its high attenuation, our paper illustrates that RF communications from air to fresh water are possible. Finally, this work derives the optimum frequencies, which provide minimum attenuation and maximum propagation depth, for RF communications from air to fresh water.

Keywords: Wireless communications, Propagation, Fresh water, Optimum frequency

1. INTRODUCTION

Underwater communications have attracted significant interest in recent years since they have a wide range of applications including coastline protection, underwater environmental observation for exploration, off-shore oil/ gas field monitoring, oceanographic data collection, autonomous underwater vehicles (AUVs) and remotely operated vehicles (ROVs), etc., [1]. Reliable data monitoring and transmission in shallow fresh water (e.g., rivers and lakes) have also received growing interest since they can provide information that is crucial for the local economies as well as the environment. Traditionally, underwater communications have been done through acoustic and optical systems that have certain advantages and disadvantages. For example, acoustic communications are widely used in underwater environment but their performance in shallow waters is severely affected by multipath propagation. Also, the channel latency caused by the low propagation velocity in water is another limitation of acoustic communications. On the contrary, laser based optical systems have significantly higher propagation speed than underwater acoustic waves. However, strong backscattering caused by suspended particles in water always limits the application of optical systems to very short distances.

Electromagnetic (EM) waves in the RF range can also be used for underwater wireless communication systems. The velocity of EM waves in water is more than 4 orders faster than acoustic waves so the channel latency is greatly reduced. In addition, EM waves are less sensitive than acoustic waves to reflection and refraction effects in shallow water. Moreover, suspended particles have very little impact on EM waves. Few underwater communication systems based on EM waves have been proposed before [2,3].

The primary limitation of EM wave propagation in water is the high attenuation due to the conductivity of water. For example, it has been shown in [4] that conventional RF propagation works poorly in seawater due to the losses caused by the high conductivity of seawater (typically, 4 S/m). However, fresh water has a typical conductivity of only 0.01 S/m, which is 400 times less than the typical conductivity of seawater. Therefore, EM wave propagation can be more efficient in fresh water than in seawater.

This work focuses on the analysis of electromagnetic waves penetrating from air to fresh water. The following two types of losses are analyzed: 1) the transmission loss that is due to reflection at the air-water interface, and 2) the propagation loss inside the water due to its material properties. The electromagnetic properties of fresh water are modeled using the Debye model. Also, the effects of the incidence angle as well as the polarization are examined and an optimum frequency range is identified for RF air-to-water communications to minimize the power loss.

2. PLANE WAVE MODEL

In this section, we model a plane wave impinging on an air-water interface assuming the depth of water is infinite (see Figure 1). For such a plane wave penetrating the fresh water, the total power loss is the sum of the transmission loss and propagation loss. Our analytical formulations describe both loss mechanisms.

2.1. Power Attenuation for Normal Incidence

We formulate analytical equations to calculate the transmission loss and propagation loss for the scenario of Figure 1. The incident power is written as:

$$P_i = \mathrm{Re}\left\{\frac{|E_i|^2}{2\eta_0}\right\} \tag{1}$$

where E_i is the incident electric field and $\eta_0 = \sqrt{\frac{\mu_0}{\varepsilon_0}}$

stands for the intrinsic impedance of air. Similarly, the transmitted power in water is written as:

$$P_t = \mathrm{Re}\left\{\frac{|E_t|^2}{2\eta_1^*}\right\} \tag{2}$$

where E_t is the transmitted electric field, and

$\eta_1 = \sqrt{\frac{\mu_1}{\varepsilon_0 \bar{\varepsilon}_r}}$ is the intrinsic impedance of water, where

$\bar{\varepsilon}_r = \varepsilon_r' - j\varepsilon_r''$ stands for the relative complex permittivity of fresh water [5].

The complex frequency dependent dielectric permittivity $\bar{\varepsilon}_r(f)$ of water is commonly described with the Debye model as:

$$\bar{\varepsilon}_r(f) = \varepsilon_\infty + \left(\frac{\varepsilon_s - \varepsilon_\infty}{1 + \left(i \frac{f}{f_{ref}} \right)} \right) - \frac{i\sigma}{2\pi f \varepsilon_0} \tag{3}$$

where ε_s and ε_∞ are the real relative permittivity at low and high frequencies, respectively, f_{ref} is the relaxation frequency, σ is the conductivity of water, and ε_0 is the dielectric permittivity of free space. Table 1 shows the Debye parameters of water reported in [5].

Further, the transmitted power in (2) can also be expressed as follows:

$$P_t = \mathrm{Re}\left\{ \frac{|E_i|^2}{2\eta_1^*} \right\} |T|^2 \, \mathrm{e}^{-2\alpha d} \tag{4}$$

where T is the transmission coefficient given by:

$$T = \frac{E_t}{E_i} = \frac{2\eta_1}{\eta_0 + \eta_1} \tag{5}$$

Table 1. Debye parameters of water [5].

Medium	ε_s, [ ]	ε_∞, [ ]	f_{ref}, [GHz]	σ, S·m^{-1}
Water	80	4.22	17.4	17.4

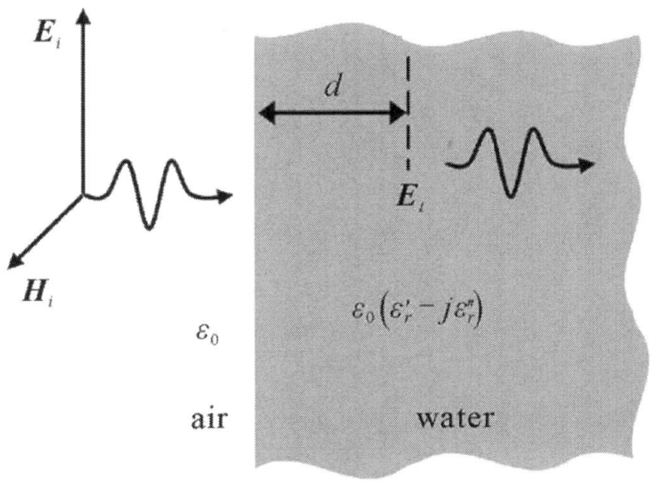

Figure 1. Plane wave penetrating fresh water at normal incidence.

Then, the transmission loss describing the power loss caused by the wave reflection at the air-water interface can be calculated in dB as follows:

$$\alpha_t = 10\log_{10}\left(|T|^2 \operatorname{Re}\left(\frac{\eta_0}{\eta_1^*}\right)\right) \tag{6}$$

When calculating the propagation loss inside water, the attenuation α [6], is given by (7),

$$\alpha = \omega\sqrt{\mu\varepsilon}\left\{\frac{1}{2}\left[\sqrt{1+\left(\frac{\sigma}{\omega\varepsilon}\right)^2}-1\right]\right\}^{\frac{1}{2}} \tag{7}$$

and the propagation loss in dB is defined as shown below:

$$\alpha_p = 10\log_{10}\left(e^{-2\alpha d}\right) \tag{8}$$

where d is the depth of propagation inside water.

Therefore, the total loss of the normal incidence is written in (9),

$$\alpha_{total} = \alpha_t + \alpha_p = 10\log\left(\frac{P_t}{P_i}\right) \tag{9}$$

and it depends on the complex permittivity of water and the depth of propagation.

2.2. Power Attenuation for Oblique Incidence

To examine transmissions at oblique incidence angles for a general wave polarization, it is convenient to decompose the electric field into perpendicular and parallel components and analyze them separately. The transmission coefficient for parallel and perpendicular polarization is given by:

$$T_{//} = \frac{2\eta_1 \cos\theta_i}{\eta_0 \cos\theta_i + \eta_1 \cos\theta_t} \tag{10}$$

$$T_{\perp} = \frac{2\eta_1 \cos\theta_i}{\eta_1 \cos\theta_i + \eta_0 \cos\theta_t} \tag{11}$$

where θ_i is the angle of incidence and θ_t is the angle of transmission, given by Snell's law of refraction [6] as

$$\cos\theta_t = \sqrt{1-\sin^2\theta_i/\bar{\varepsilon}_r}\,.$$

The transmission loss for oblique incidence can be written as follows using (6) and (10)-(11):

$$\alpha_{t//} = 10\log_{10}\left(\left|T_{//}\right|^2 \mathrm{Re}\left(\frac{\eta_0}{\eta_1^*}\right)\right) \tag{12}$$

$$\alpha_{t\perp} = 10\log_{10}\left(\left|T_{\perp}\right|^{2} \mathrm{Re}\left(\frac{\eta_0}{\eta_1^{*}}\right)\right) \tag{13}$$

The propagation loss for oblique incidence is written as:

$$\alpha_{p//} = \alpha_{p\perp} = 10\ \log_{10}\left(\mathrm{e}^{-2\alpha d/\cos(\theta_t)}\right) \tag{14}$$

Therefore, the total loss of the oblique incidence is given by

$$\alpha_{total//} = \alpha_{t//} + \alpha_{p//} \tag{15}$$

$$\alpha_{total\perp} = \alpha_{t\perp} + \alpha_{p\perp} \tag{16}$$

and it depends on the complex permittivity of water as well as the angle of incidence.

3. RESULTS

The transmission loss as well as propagation loss for a plane wave propagating from air to water is analyzed in the frequency range of 23 kHz to 1 GHz for normal incidence. This frequency range includes frequencies from the VLF (used in submarine communications) to the UHF band. Various propagation depths are considered since the propagation loss increases as the depth increases. In addition, the transmission loss remains the same for all propagation depths.

Figure 2 illustrates that the transmission loss decreases dramatically as the frequency increases from 23 kHz to 10 MHz, and then remains almost constant for frequencies higher than 10 MHz.

The propagation loss (see Figure 3) exhibits an opposite trend when compared to the trend of the transmission loss. Specifically, Figure 3(a) shows that it increases slowly for frequencies up to 100 MHz and then increases dramatically for higher frequencies (especially for frequencies around 1 GHz).

Furthermore, the transmission loss and propagation loss are added together to obtain the total power loss for

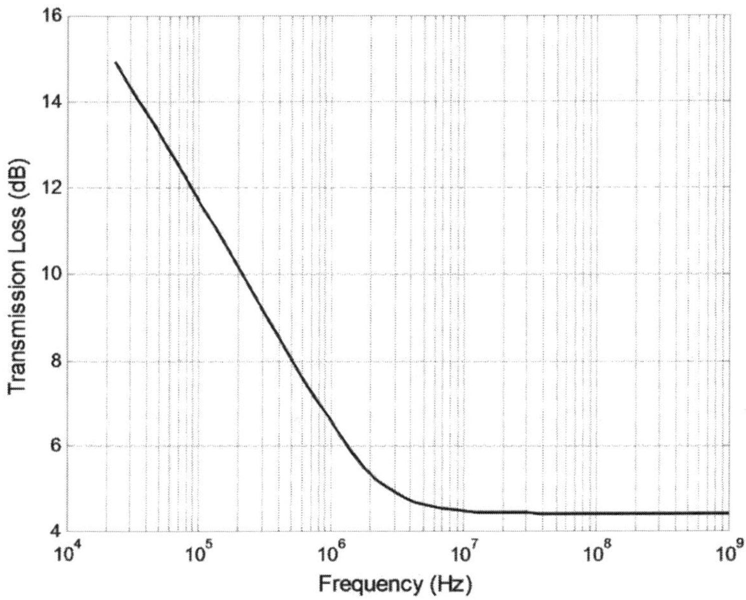

Figure 2. Transmission loss for normal incidence.

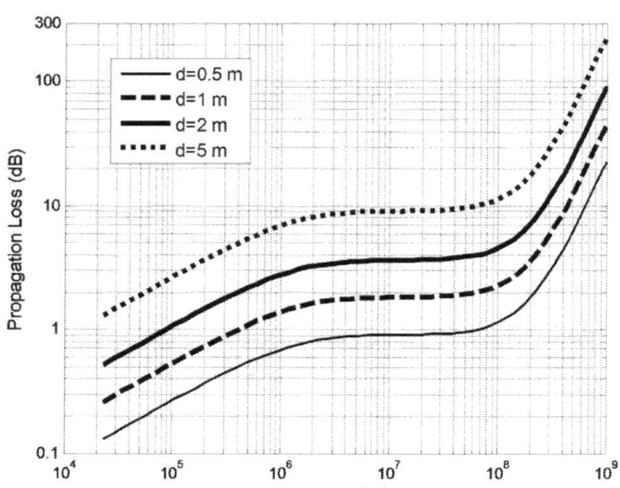

(a)

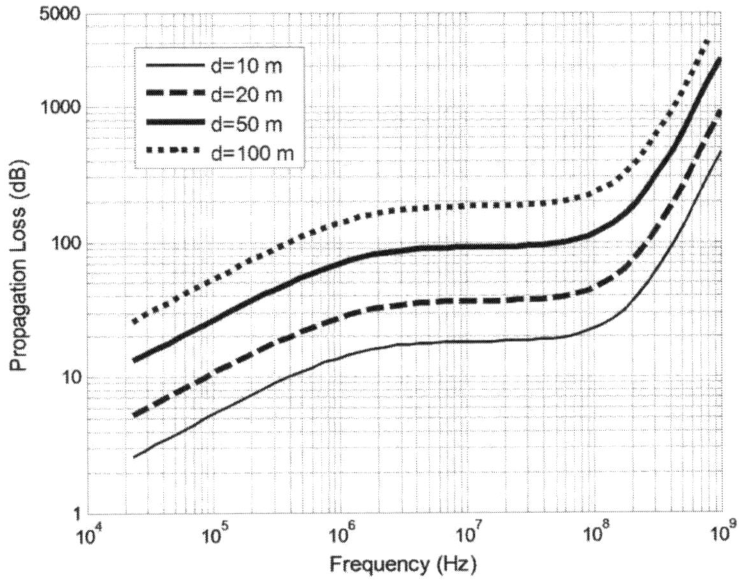

(b)

Figure 3. Propagation loss for normal incidence at different propagation depths, d: (a) shallow and (b) deep.

the air-to-water propagation (see Figure 1) and are plotted in Figure 4. As expected, due to the reverse variations of the transmission and propagation losses, an optimum frequency range exists for shallow propagation depths. In this optimum frequency range, there is significantly smaller power loss. For example, in Figure 4(a), the total loss in 3 - 100 MHz frequency range for a depth of 1 m is 10 dB to 45 dB smaller than the loss at the lowest and highest frequencies of our frequency range. Therefore, according to Figure 4(a) there is a range of optimum frequencies that exhibit minimum losses when a wave propagates from air to shallow fresh water depths. This range of frequencies can improve RF communications with underwater vehicles, or devices. Potential applications that can benefit from operating in the optimum frequency range include the following: 1)

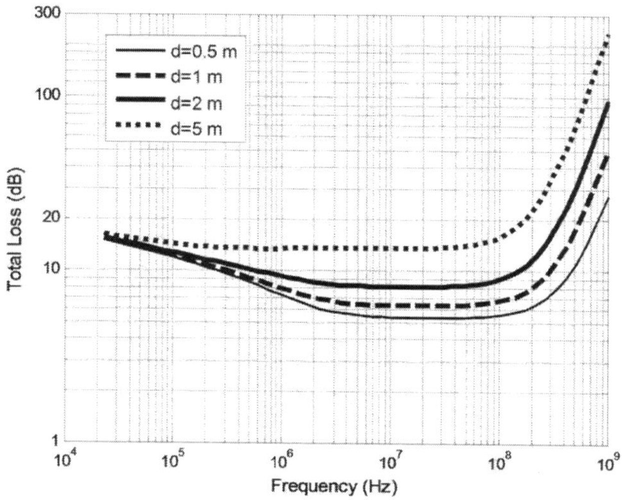

(a)

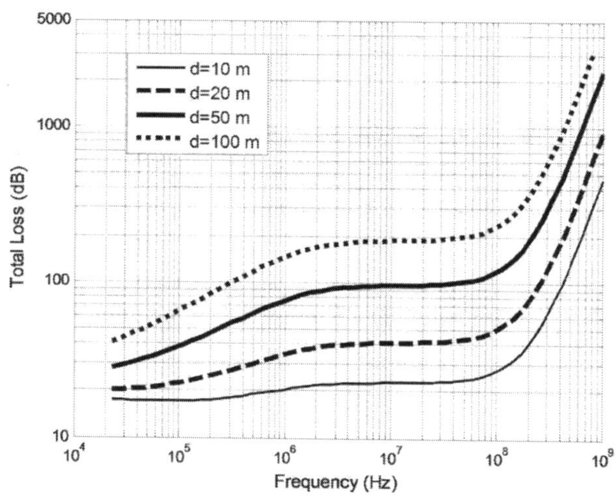

(b)

Figure 4. Total attenuation for normal incidence at different propagation depths, d: (a) shallow and (b) deep.

communications with underwater robots [7], 2) RFID based shore erosion detection [8], 3) water quality monitoring of bodies of fresh water (e.g., temperature, pH, etc.), such as, artificial lakes, swimming pools, and water tanks, using wireless underwater sensors.

Figure 4(b) also illustrates that the total loss monotonically increases when the propagation depth is larger than 10 m, thereby providing no optimum frequency range. This happens because the transmission loss is the same for different propagation depths whereas the propagation loss increases as the propagation depth becomes larger. Therefore, as the depth increases it reaches a value for which the propagation loss becomes larger than the transmission loss. Therefore, the optimum frequency range exists only for small propagation depths (less than 5 m).

Following similar steps to the normal incidence case of water half-space, the total attenuation for oblique incidence is calculated across incident angles from 0 to 89 degrees for both parallel and perpendicular polarization cases. Figure 5 illustrates that normal incidence has the least power loss for fixed frequency and propagation depth, since the total attenuation increases along with the incident angle for both polarization cases. For example, for parallel polarization, the total loss increases slowly for angles up to 80 degrees, but increases dramatically for larger incident angles.

Also, the average total loss across all incident angles ranging from 0 to 89 degrees is calculated for different propagation depths. Figure 6 plots the average total loss versus frequency for both parallel and perpendicular polarization cases. It can be concluded that the average total loss for an oblique incident plane wave is larger than the normal incidence case, but with similar trend. For example, for a propagation depth of 1 m, the average total loss

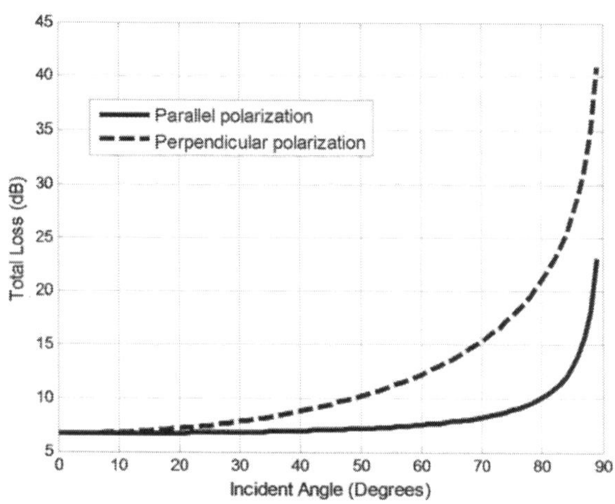

Figure 5. Total attenuation for oblique incidence when f = 100 MHz, d = 1 m.

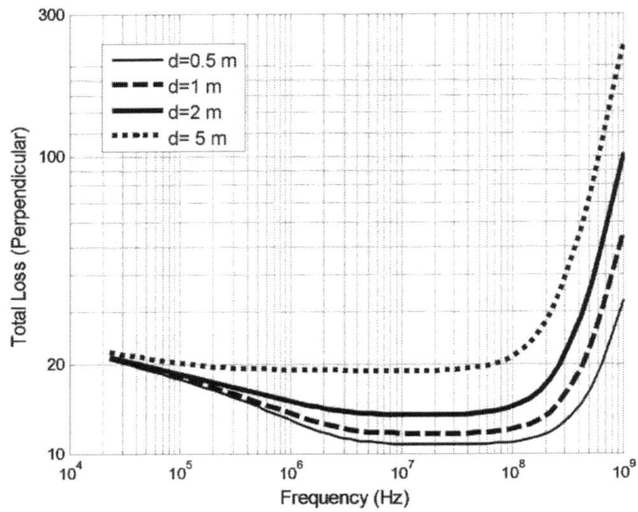

(a)

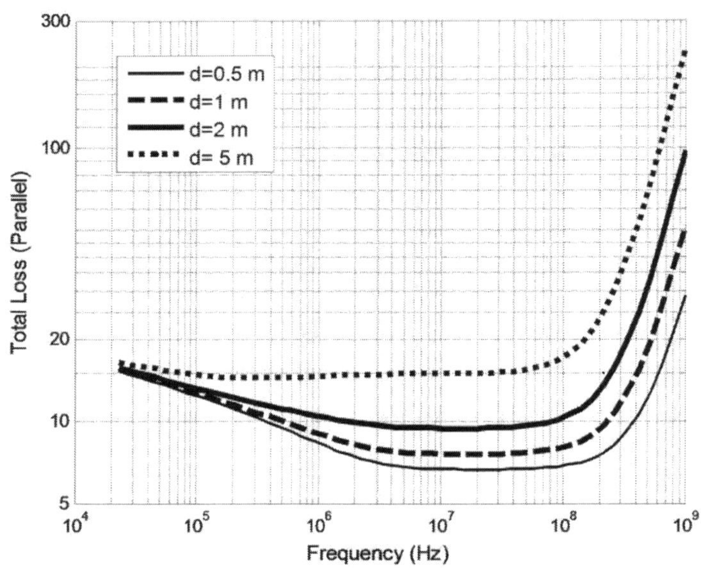

(b)

Figure 6. Average total loss across all incidence angles from 0 to 89 degrees at different depths for: (a) parallel polarization and (b) perpendicular polarization.

for perpendicular polarization (see Figure 6(b)) in the 3 MHz to 100 MHz frequency range is about 5 dB more than the total loss for normal incidence (see Figure 4(a)). From the results of Figure 6 for the oblique incidence case we observe that there is also an optimum frequency range similar to the one we identified for the normal incidence case.

4. CONCLUSIONS

A plane wave model for the air-to-water interface has been used to calculate the transmission loss, propagation loss as well as the total loss at various incident angles. The values of these losses depend on the electromagnetic properties of water, frequency, incidence angle and the propagation depth. For such air-to-water communications, an optimum frequency range (3 - 100 MHz) was identified when the plane wave propagates to depths less than 5 m. In this optimum frequency range, the wave experiences significantly smaller losses than the losses at the lowest and highest frequencies of our analysis. Specifically, this frequency range includes the bands of shortwave radio (3 - 30 MHz), VHF TV (54 - 72 MHz, 76 - 88 MHz), parts of FM (88 - 108 MHz) and US military VHF-FM (30 - 88 MHz). Therefore, various communications systems can benefit from using the optimum operation frequencies that we identified here. Also, wireless power harvesting by wireless sensors can be significantly enhanced if it is performed inside the 3 - 100 MHz range.

It should be also pointed out, that practical compact antenna designs can be developed in the optimum frequency range of 3 - 100 MHz for underwater devices due to the large permittivity of water ($\varepsilon_r = 81$). For example, a half-wavelength loop antenna operating at 100 MHz inside fresh water has a diameter of only 5.3 cm. Therefore, the optimum frequency range can be used in practical underwater communication systems and it will provide minimum losses for shallow propagation depths.

5. ACKNOWLEDGEMENTS

This work was supported by the Dissertation Year Fellowship provided by Florida International University's Graduate School.

REFERENCES

1. L. Liu, S. Zhou and J. Cui, "Prospects and Problems of Wireless Communication for Underwater Sensor Networks," Wiley WCMC Special Issue on Underwater Sensor Networks (Invited), 2008.

2. J. H. Goh, A. Shaw, A. I. Al-Shanmma'a, "Underwater Wireless Communication System," Journal of Physics, Conference Series 178, 2009.

3. A. I. Al-Shamma'a, A. Shaw and S. Saman, "Propagation of Electromagnetic Waves at MHz Frequencies through Seawater," IEEE Transactions on Antennas and Propation, Vol. 52, No. 11, November 2004, pp. 2843-2849.

4. S. Bogie, "Conduction and Magnetic Signaling in the Sea," Radio Electronic Engineering, Vol. 42, No. 10, 1972, pp. 447-452.

5. J. B. Hasted, "Aqueous Dielectrics," Chapman and Hall, New York, 1973.

6. C. A. Balanis, "Advanced Engineering Electromagnetics," John Wiley & Sons, New York, 1989.

7. T. Shaneyfelt, M. A. Joordens, K. Nagothu and M. Jamshidi, "RF Communication between Surface and Underwater Robotic Swarms," Automation Congress, Hawaii, 9 December 2008.

8. G. Benelli, A. Pozzebon and G. Ragseo, "An RFID Based System for the Underwater Tracking of Pebbles on Artificial Coarse Beaches," IEEE 2009 3rd International Conference on Sensor Technologies and Applications, Athens, 18-23 June 2009, pp. 294-200.

3
CHAPTER
LATERAL WAVES NEAR THE SURFACE OF SEA

Osama M. Abo-Seida[1], Samira T. Bishay[2], Khaled M. El-Morabie[3]

[1]*Department of Mathematics, Faculty of Science, Kafer El-Sheikh University, Kafer El-Sheikh, Egypt*

[2]*Department of Mathematics, Faculty of Science, Ain Shams University, Abbassia, Cairo, Egypt*

[3]*Department of Mathematics, Faculty of Science, Tanta University, Tanta, Egypt.*

ABSTRACT

In this research, we investigate the propagation of lateral electromagnetic wave near the surface of sea. Interference patterns generated by the superposition of the lateral and direct waves along the sea surface (flat and rough) are shown. The field generated by a vertical magnetic dipole embedded below the sea surface (having a flat and perturbed upper surface) is shown to consist of a lateral-wave and a reflected-wave. Closed-form expressions for the lateral waves near the surface of the sea are obtained and compared with those mentioned for the reflected waves numerically for the considered model.

Keywords: Stratified Media, Rough Surface, Radiation in Sea, Lateral Waves

1. INTRODUCTION

Lateral electromagnetic waves generated by a vertical electric or magnetic dipole near the plane boundary between two different media like air and earth or air and sea have been the subject of investigation for many years beginning with the work of Sommerfeld. King [1] derived simple formulas for the transient field generated by a vertical electric dipole on the boundary between two dielectric half-space when the permittivity of one of these is much greater than that of the other. The roughness of the upper surface of the sea is considered by Bishay [2,3] to indicate the effect of the rough surface on the electromagnetic fields. Recently, Abo-Seida et al. [4] calculated the far-field radiated from a vertical magnetic dipole in sea with a rough upper surface. Besides, in previous studies [4], the Hankel transformations are estimated by using new technique developed by Long et al. [5] and Chew [6].

The present study is a further contribution to [4], so the Hankel transformations which were estimated by Abo-Seida et al. [4] are employed here. The previous studies [2,3] have obtained the formulas of the reflected waves in the region of the seawater, due to a vertical magnetic dipole in a three-layered conducting media by resolving the problem using the residue and saddle-point methods.

However, these methods, involve lengthy algebra and several transformations, which are very tedious and complicated. The new technique utilized in [4] was used in this study in order to obtain closed-form expressions of the lateral waves.

Firstly the form solutions of the far-field, due to a vertical magnetic dipole in a sea (three-layered conducting media) with variable interface are expanded as an infinite series. Then with the aid of the complex image theory [7], closed-form expression of the lateral waves near the sea surface due to the dipole are obtained. Besides, the physical meaning of the results is presented.

2. GEOMETRICAL STRUCTURE

We shall adopt the following model as illustrated in Figure 1. A small loop antenna, whose magnetic moment is IS_0, is located in the middle layer (i.e. in the sea) at depth d_1, horizontally. An observing point P is also located in the sea at depth d_2. We suppose that the thickness of the sea is a, and that air is infinite upward and the ground downward along the z-axis. The groundsea interface is taken to be planar, while the air-sea interface varies slightly from its mean value, as shown in the figure. r is the distance between the source and the observing point P, and r_1 is that between P and the image of the source in the air.

The material constants are assumed to be as follows. The dielectric constant in the air, sea, and ground are ε_0, ε_2 and ε_3, respectively. The magnetic permeability is taken equal to that of the free space in every layer. The conductivity of the seawater and ground are σ_2 and σ_3, respectively.

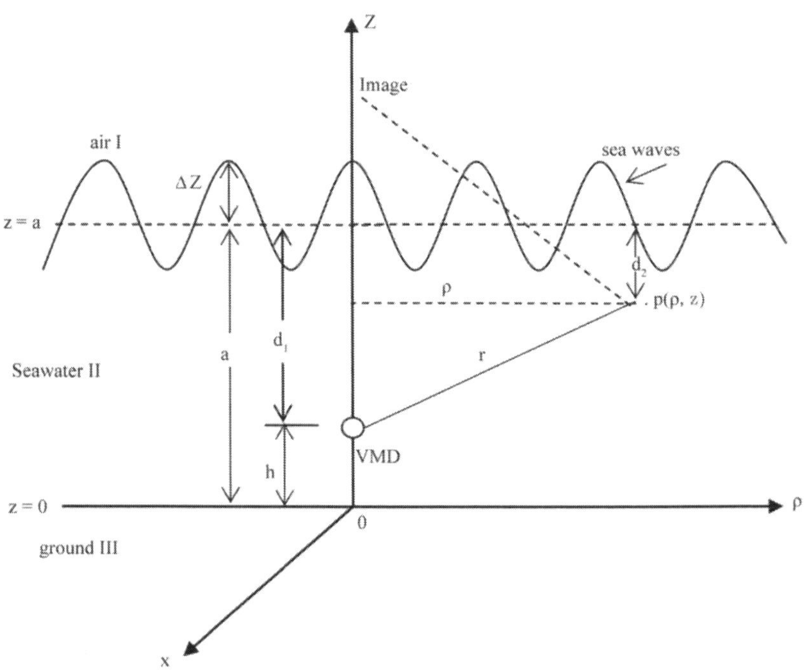

Figure 1. Geometrical configuration of the problem.

3. The Lateral Waves in Sea

Lateral waves are the electromagnetic waves which are generated by vertical or horizontal dipoles on/or near the plane boundary between two electrically different media like air and earth or air and sea or ocean water. However, the lateral wave propagates from the antenna to the surface suffering some attenuation, then propagates in the air without attenuation over a long distance and then arrives at the receiver. Thus the communication is only through the lateral wave, since the direct and reflected waves are almost completely attenuated in the medium as in this case of flat upper surface.

To evaluate the lateral waves, we return to the Equation (29) in [4] as

$$E^2_{so\varphi} = \frac{-j\omega\mu_0 IS_0}{8\pi}\left[\int_{-\infty}^{\infty}\frac{\lambda}{u_2}\frac{\left(\frac{u^-_{21}}{u^+_{21}}\right)e^{-2u_2a}e^{u_2h}}{[1-\left(\frac{u^-_{21}u^-_{23}}{u^+_{21}u^+_{23}}\right)e^{-2u_2a}]}e^{u_2z}H^{(2)}_1(\lambda\rho)\lambda d\lambda + \int_{-\infty}^{\infty}\frac{\lambda}{u_2}\frac{\left(\frac{u^-_{21}u^-_{23}}{u^+_{21}u^+_{23}}\right)e^{-2u_2a}e^{-u_2h}}{\left[1-\left(\frac{u^-_{21}u^-_{23}}{u^+_{21}u^+_{23}}\right)e^{-2u_2a}\right]}e^{u_2z}H^{(2)}_1(\lambda\rho)\lambda d\lambda\right]$$

where $u_i = \sqrt{\lambda^2 - k_i^2}$, $(i = 1,2,3)$ and its real part is positive, k_i is the propagation constant of the medium under consideration. $u^{\pm}_{ij} = (u_i \pm u_j)$, $(i, j = 1,2,3)$ and $H^{(2)}_1(\lambda\rho)$ is the second-kind Hankel functions of order one. Therefore, we can write $E^2_{so\varphi}$ as

$$E^2_{so\varphi} = \frac{-j\omega\mu_0 IS_0}{8\pi}\left\{\sum_{m=0}^{\infty}\phi(m) + \sum_{m=0}^{\infty}\psi(m)\right\} \tag{1}$$

where

$$\phi(m) = \int_{-\infty}^{\infty}\frac{\lambda^2}{u_2}\left(\frac{u^-_{21}u^-_{23}}{u^+_{21}u^+_{23}}\right)^m\left(\frac{u^-_{21}}{u^+_{21}}\right)e^{-u_2[d_1+d_2+2ma]}H^{(2)}_1(\lambda\rho)d\lambda \tag{2}$$

$$\psi(m) = \int_{-\infty}^{\infty}\frac{\lambda^2}{u_2}\left(\frac{u^-_{21}u^-_{23}}{u^+_{21}u^+_{23}}\right)^{m+1} e^{-u_2[d_1+d_2++2h+2ma]}H^{(2)}_1(\lambda\rho)d\lambda \tag{3}$$

According to the complex image theory [7], when condition $|k_1^2| << |k_2^2|$ is satisfied, we have $\frac{u_{21}^-}{u_{21}^+} \approx e^{-u_1 d}$,

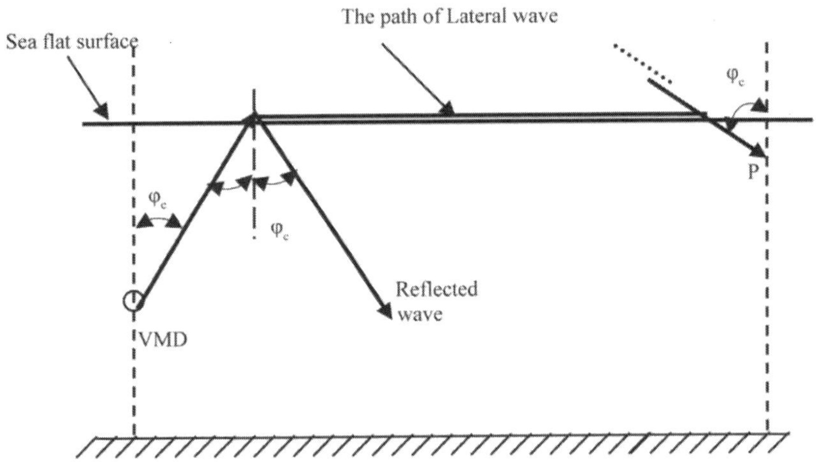

Figure 2. The paths of reflected and lateral waves.

and

$$\frac{2u_1}{u_{21}^+} \approx 1 - e^{-u_1 d} \tag{4}$$

where $d = \frac{-2j}{k_2}$.

Using the first formula of (4) and rearranging properly $\phi(m)$ and $\psi(m)$ to obtain the lateral waves. Then, we have

$$(E_{so\varphi}^2)_{lat} = \frac{-j\omega\mu_0 IS_0}{8\pi}\left\{\sum_{m=0}^{\infty}\varphi_{lat}(m) + \sum_{m=0}^{\infty}\psi_{lat}(m)\right\} \tag{5}$$

where

$$\phi_{lat}(m)=\int_{-\infty}^{\infty}\left\{\lambda\frac{u_1}{u_2}\left(\frac{u_{23}^-}{u_{23}^+}\right)^m e^{-u_2(d_1+d_2+2ma)}\frac{H_1^{(2)}(\lambda\rho)}{H_0^{(2)}(\lambda\rho)}\right\}\left\{\frac{\lambda}{u_1}e^{-u_1(m+1)}H_0^{(2)}(\lambda\rho)\right\}d\lambda \tag{6}$$

$$\psi_{lat}(m)=\int_{-\infty}^{\infty}\left\{\lambda\frac{u_1}{u_2}\left(\frac{u_{23}^-}{u_{23}^+}\right)^{m+1} e^{-u_2(d_1+d_2+2h+2ma)}\frac{H_1^{(2)}(\lambda\rho)}{H_0^{(2)}(\lambda\rho)}\right\}\left\{\frac{\lambda}{u_1}e^{-u_1(m+1)}H_0^{(2)}(\lambda\rho)\right\}d\lambda \tag{7}$$

The first factor in (6) is a slowly varying part, while the second term is rapidly varying. The location of the stationary phase point is given by

$$\frac{\partial}{\partial\lambda}j\left[-\lambda\rho-\sqrt{k_1^2-\lambda^2}\,(m+1)d\right]\Bigg|_{\lambda=\lambda_0}=0 \tag{8}$$

The solution of (8) is

$$\lambda_0=\frac{k_1\rho}{r_{m+1}} \tag{9}$$

where, $r_{m+1}=\sqrt{\rho^2+(m+1)^2d^2}$ We can see that the approximation in (9) is valid for $\lambda\simeq k_1$

$u_1=\frac{jk_1}{r_{m+1}}(m+1)d$ and $u_2=jk_2$, then we get

$$\phi_{lat}(m)=\frac{4k_1^2}{k_2^2}\rho(m+1)e^{-jk_2(2ma+d_1+d_2)}\left(\frac{1-n_3}{1+n_3}\right)^m\frac{e^{-jk_1r_{m+1}}}{r_{m+1}^3} \tag{10}$$

where $n_3=\frac{k_3}{k_2}$

In analogy to this approach we can obtain $\psi_{lat}(m)$ as follows

$$\psi_{lat}(m)=\frac{4k_1^2}{k_2^2}\rho(m+1)e^{-jk_2(2ma+2h+d_1+d_2)}\left(\frac{1-n_3}{1+n_3}\right)^{m+1}\frac{e^{-jk_1r_{m+1}}}{r_{m+1}^3} \quad (11)$$

From (10) and (11), we get

$$(E_{so\varphi}^2)_{lat}=\frac{-j\omega\mu_0 IS_0}{2\pi}\frac{k_1^2}{k_2^2\rho^2}e^{-jk_1\rho-jk_2(d_1+d_2)}\left[\begin{array}{l}\sum_{m=0}^{\infty}(m+1)\left(\frac{1-n_3}{1+n_3}\right)^{m}e^{-jk_2 2ma}\\+\sum_{m=0}^{\infty}(m+1)\left(\frac{1-n_3}{1+n_3}\right)^{m+1}e^{-jk_2(2h+2ma)}\end{array}\right] \quad (12)$$

The exponential in the lateral wave term all indicates that the wave travel vertically a distance d_1+d_2 from the dipole to the boundary surface in the sea, then horizontally along the boundary a distance ρ in the air, and finally vertically in the sea to the point of observation.

4. LATERAL WAVES AT THE ROUGH SURFACE OF THE SEA

In this section we shall calculate the lateral waves for secondary fields in the sea. These will represent the changes which occur in the electromagnetic field of the wave propagation in the sea due to the perturbation applied on the upper surface, where

$$E_{sg}^2=\begin{cases}\left(E_{sg}^2\right)_{ref}, & \text{when waves with propagation constant } k_2\\ \left(E_{sg}^2\right)_{lat}, & \text{when waves with propagation constant } k_1\end{cases},\quad g=0,1 \quad (13)$$

$(E_{sg}^2)_{ref}$: The secondary reflected electric field in the sea in any direction,

$(E_{sg}^2)_{lat}$: The secondary lateral electric field in the sea in any direction, as in Figure 3.

The reflected field $\left(E_{s1\varphi}^2\right)_{ref}$ has been calculated in the previous paper [4]. We can calculate $\left(E_{s1\varphi}^2\right)_{lat}$ by using the following approach, in which the denominator can be expanded as an infinite series, i.e.,

$$\mathrm{E}_{s1\varphi}^{2} = \frac{-j\omega\mu_0 IS_0}{8\pi}\int_{-\infty}^{\infty} B_1(\lambda)e^{u_2 Z}H_1^{(2)}(\lambda\rho)\lambda \mathrm{d}\lambda \tag{14}$$

Then,

$$E_{s1\varphi}^{2} = \frac{-j\omega\mu_0 IS_0}{8\pi}\left\{\sum_{n=0}^{\infty}\phi(n)+\sum_{n=0}^{\infty}\psi(n)\right\} = E_1 + E_2 \tag{15}$$

whereIn this research, there is always $\left|k_2^2\right| \geq 80\left|k_1^2\right|$ for any frequency, therefore, the condition required for (4) is met. Using the first formula of (4) and arranging properly, we get

$$\phi(n) = \int_{-\infty}^{\infty}\lambda\left(\frac{u_1}{u_{21}^{+}}\right)\left(\frac{u_{21}^{-}u_{23}^{-}}{u_{21}^{+}u_{23}^{+}}\right)^{n} e^{-2u_2 na}\left[M_0(\lambda)+\frac{j}{u_1}L_0(\lambda)\right]e^{-u_2(d_1+d_2)}H_1^{(2)}(\lambda\rho)\mathrm{d}\lambda, \tag{16}$$

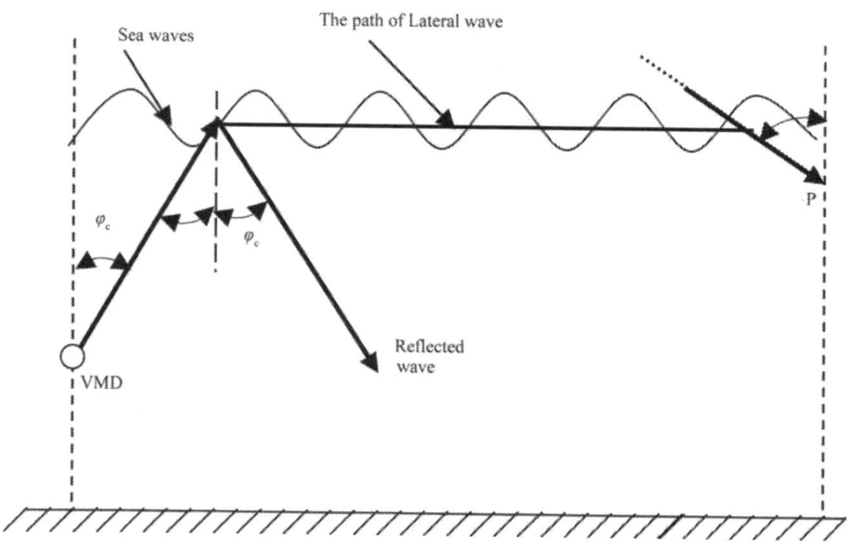

Figure 3. The paths of reflected and lateral waves.

$$\psi(n)=\int_{-\infty}^{\infty}\lambda\left(\frac{u_1}{u_{21}^{+}}\right)\left(\frac{u_{21}^{-}u_{23}^{-}}{u_{21}^{+}u_{23}^{+}}\right)^{n}e^{-u_2(2na+d_2)}\left[M_m(\lambda)+\frac{j}{u_1}L_m(\lambda)\right]e^{-jk_2d_1}H_1^{(2)}(\lambda\rho)\mathrm{d}\lambda \tag{17}$$

$$\begin{aligned}\phi_{lat}(n)&=I_{1lat}+I_{2lat}\\&=\int_{-\infty}^{\infty}\left\{\frac{\lambda}{u_1}e^{-u_1nd}H_0^{(2)}(\lambda\rho)\right\}\left\{u_1\left(\frac{u_{23}^{-}}{u_{23}^{+}}\right)^{n}e^{-u_2(2na+d_1+d_2)}\left[M_0(\lambda)+\frac{j}{u_1}L_0(\lambda)\right]\frac{H_1^{(2)}(\lambda\rho)}{H_0^{(2)}(\lambda\rho)}\right\}\mathrm{d}\lambda\\&\quad-\int_{-\infty}^{\infty}\left\{\frac{\lambda}{u_1}e^{-u_1(n+1)d}H_0^{(2)}(\lambda\rho)\right\}\left\{u_1\left(\frac{u_{23}^{-}}{u_{23}^{+}}\right)^{n}e^{-u_2(2na+d_1+d_2)}\left[M_0(\lambda)+\frac{j}{u_1}L_0(\lambda)\right]\frac{H_1^{(2)}(\lambda\rho)}{H_0^{(2)}(\lambda\rho)}\right\}\mathrm{d}\lambda\end{aligned} \tag{18}$$

When $\lambda\rho\to\infty$, there is an approximate formula for Hankel function as

$$H_n^{(2)}(\lambda\rho)\cong\sqrt{\frac{2}{\pi\lambda\rho}}e^{-j(\lambda\rho-n\pi/2-\pi/4)}.$$

Then, when $\lambda\rho\to\infty$, the first bracket in I_1 is slowly varying part while the second bracket is rapidly varying. The location of the stationary phase point in I_1 is given by:

$$\frac{\partial}{\partial\lambda}j\left[-\lambda\rho-\sqrt{k_1^2-\lambda^2}\,(nd)\right]\Big|_{\lambda=\lambda_0}=0 \tag{19}$$

The solution of (19) is given by

$$\lambda_0=\frac{k_1\rho}{r_n} \tag{20}$$

where, $r_n=\sqrt{\rho^2+n^2d^2}$. We can see that the approximation in (4) is valid for $\lambda\cong k_1$. Apparently the reason of this validity is because the stationary phase point ends up being at $\lambda\cong k_1$.

Using the approach we presented in the pervious paper [4], we get

$$I_{1lat} = \left\{ u_1 \left(\frac{u_{23}^-}{u_{23}^+}\right)^n e^{-u_2(2na+d_1+d_2)} \left[M_0(\lambda) + \frac{j}{u_1} L_0(\lambda) \right] \frac{H_1^{(2)}(\lambda\rho)}{H_0^{(2)}(\lambda\rho)} \right\}_{\lambda=\lambda_0} \cdot \frac{e^{jk_1 r_n}}{r_n}$$
$$= \left\{ k_1^2 e^{-jk_2(2na+d_1+d_2)} \left(\frac{1-n_3}{1+n_3}\right)^n \left[\bar{M}_0(\lambda) + \frac{j}{u_1} \bar{L}_0(\lambda) \right] \right\} \frac{nd\rho e^{-jk_1 r_n}}{r_n^3} \tag{21}$$

Similarly,

$$I_{2lat} = \left\{ k_1^2 e^{-jk_2(2(n+1)a+d_1+d_2)} \left(\frac{1-n_3}{1+n_3}\right)^n \left[\bar{M}_0(\lambda) + \frac{j}{u_1} \bar{L}_0(\lambda) \right] \right\} \frac{(n+1)d\rho}{r_{n+1}^3} e^{jk_1 r_{n+1}} \tag{22}$$

Then, from (21) and (22) we get

$$\phi_{lat}(n) = \left\{ k_1^2 e^{-jk_2(2na+d_1+d_2)} \left(\frac{1-n_3}{1+n_3}\right)^n \left[\bar{M}_0(\lambda) + \frac{j}{u_1} \bar{L}_0(\lambda) \right] \right\} \left\{ \frac{nd\rho e^{-jk_1 r_n}}{r_n^3} - \frac{(n+1)d\rho e^{-jk_1 r_{n+1}}}{r_{n+1}^3} \right\}$$

By using the following approximation, $\phi_{lat}(n)$ can be expressed as

$$\phi_{lat}(n) = \left\{ \frac{2jk_1^2}{k_2\rho} e^{-jk_2(2na+d_1+d_2)} \left(\frac{1-n_3}{1+n_3}\right)^n \left[\bar{M}_0(\lambda_0) + \frac{1}{k_1} \bar{L}_0(\lambda_0) \right] \right\} e^{-jk_1\rho} \tag{23}$$

In the same way,

$$\psi_{lat}(n) = \left\{ \frac{2jk_1^2}{k_2\rho} e^{-jk_2(2na+d_1+d_2)} \left(\frac{1-n_3}{1+n_3}\right)^n \left[\bar{M}_m(\lambda_1) + \frac{1}{k_1} \bar{L}_m(\lambda_1) \right] \right\} e^{-jk_1\rho} \tag{24}$$

Hence,

$$E_{s1\varphi}^2 = \frac{\omega\mu_0 IS_0}{4\pi} \frac{k_1^2 e^{-jk_1\rho}}{k_2\rho} e^{-jk_2(d_1+d_2)}$$
$$\times \sum_{n=0}^{\infty} e^{-jk_2 2na} \left(\frac{1-n_3}{1+n_3}\right)^n \left\{ \left[\bar{M}_m(\lambda_1) + \bar{M}_0(\lambda_0) \right] + \frac{1}{k_1} \left[\bar{L}_m(\lambda_1) + \bar{L}_0(\lambda_0) \right] \right\} \tag{25}$$

where,

$$\bar{M}_0(\lambda)=\left[\left(1-\frac{2u_2}{d_1k_2^2}\right)f(\rho_0)\right],\ \bar{L}_0(\lambda)=\frac{-j}{d_1}\left(\frac{3u_2^2}{k_2^2}-u_2d_1+1\right)f(\rho_0)-\frac{\lambda}{u_2}f'(\rho_0)$$

From (25), we have found the lateral waves near the rough surface of the sea. The physical meaning of the first term in (25) indicates a series of waves that propagate upward from the source and make n round trips between the rough sea surface and the bottom, then travel along the surface on the seawater, and finally arrive at the field point in sea, where $\left(\frac{1-n_3}{1+n_3}\right)$ is the reflection coefficient at the sea bottom, as in Figure 4.

The second term represents another series of waves that propagate downward from the source first then reflected upward at the sea bottom and travel along the same path. These results also show that $\phi_{lat}(n)$ and $\psi_{lat}(n)$ are proportional with $e^{-jk_1\rho}/\rho$, and this means that the electric field in the sea takes the form of the Sommerfeld integral. Also, the result coincides with the result obtained by Long et al. [5], when the disturbed factor $f_n(\rho)$ is neglected.

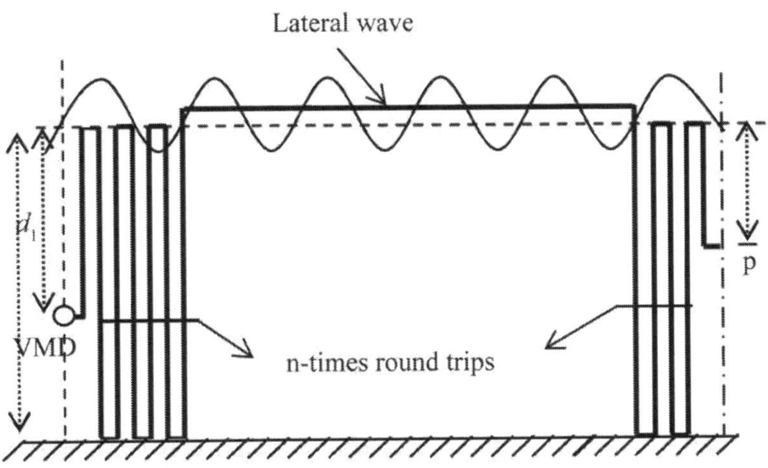

Figure 4. The propagation paths of lateral waves between the source and the bottom to receiver.

5. NUMERICAL RESULTS

The magnitudes of the electric field (the reflected and lateral waves) for the two cases (flat and rough) in sea are computed for different values of the sea thickness and different frequencies.

Case I: Flat surface of the sea: the Figures 5 to 7 show the normalized electric field of the dipole in sea versus the radial distance factor ρ. The depth d_2 of the field point is taken equal to 5 m. To attain the numerical calculation, we ascertain that $\sum_{m=0}^{25}$ is a match for $\sum_{m=0}^{\infty}$ under the conditions shown in Figures 5 to 7.

We show in these figures that increasing the depth of the sea decreases the value of the electric field. Thus, the previous equations have led to those valid and useful results. These findings also indicate the importance of the derived equations in this research area that could be easily applied in treating related problems.

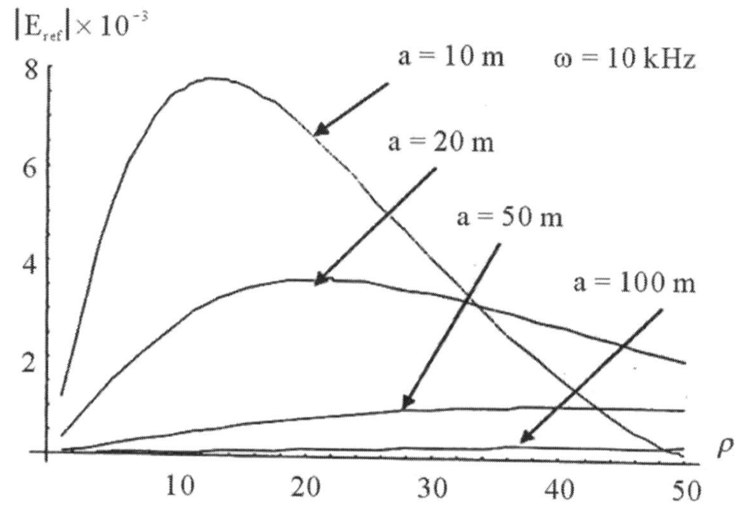

Figure 5. The reflected waves in sea with uniform surface.

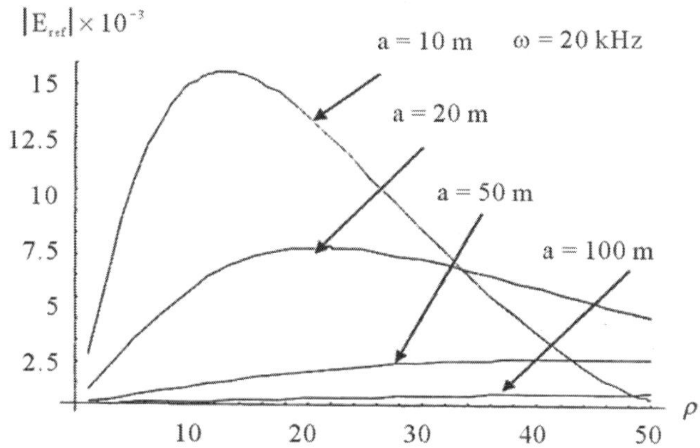

Figure 6. The reflected waves in sea with uniform surface.

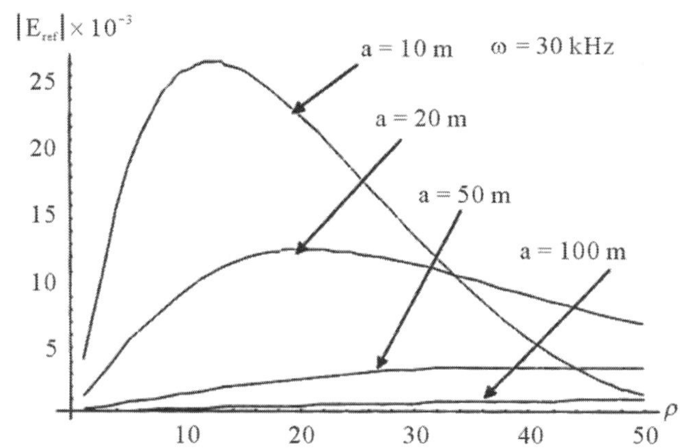

Figure 7. The reflected waves in sea with uniform surface.

Plots of the variation of the electric field in the case of the lateral waves are shown in Figures 8 to 10. Also, in these Figures, the numerical calculations ascertain that

$\sum_{m=0}^{25}$ is a match for $\sum_{m=0}^{\infty}$. Also, we show that as the depth of the sea increases, the magnitude of the electric field near the air-sea boundary surface proceeds parallel to the horizontal scale.

In the Figure 5, we consider that the horizontal axis denote the radial distance ρ and the vertical axis denote the normalized reflected electric field $|E_{ref}|$. To guarantee that our work proves correctness, we took four different values for the sea depth (a) as shown in Figure 5. The frequency (ω) was also taken to be 10 KHz. Accordingly, the results show that as we increase the sea depth, the value of the absolute reflected electric field decreases as the radial distance factor (ρ) increases.

Furthermore, we used the same values for the sea depths, but altered the values of the frequency to demonstrate that the absolute reflected electric field increases as we increase the frequency. This is clearly shown in Figures 5-7.

In addition to the reflected wave, we also examined the lateral wave. We used the same sea depth and the same frequencies. Accordingly, we got the previous three Figures 8, 9 and 10. They show that as the sea depth increases and as the frequency decreases, the lateral wave also decreases.

Case II: Rough surface of the sea. The perturbed lateral electric field $E_{S1\varphi}^2 = E_{\varphi,lat}^2$ for the rough upper surface and reflected electric field $E_{S1\varphi}^2 = E_{\varphi,ref}^2$ for the same mode are computed. The part of the sea under investigation takes different thickness (a = 10 m, 25 m, 50 m), the height of the source is taken to be 4 m from the ground (the bottom of the sea). We chose the sea waves represented by the roughness cosine profile with period

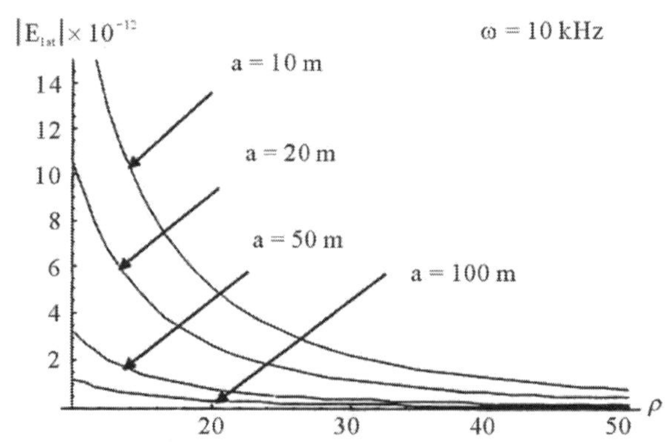

Figure 8. The lateral waves in sea with uniform surface.

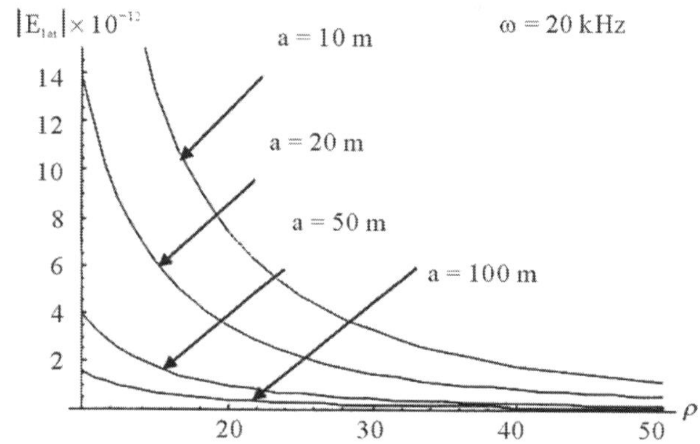

Figure 9. The lateral waves in sea with uniform surface.

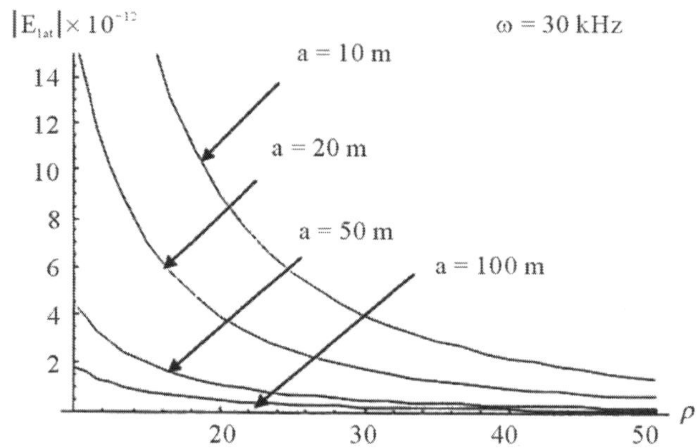

Figure 10. The lateral waves in sea with uniform surface.

10 m and for frequency 10 KHz. Consequently, Figure 11 shows that the reflected waves are decreasing rapidly with the increasing radial distance ρ at different sea depths.

Moreover, in Figure 12 we illustrate that lateral waves act appositely to the reflected waves where they are increasing hastily with the increasing radial distance ρ at the same different depths.

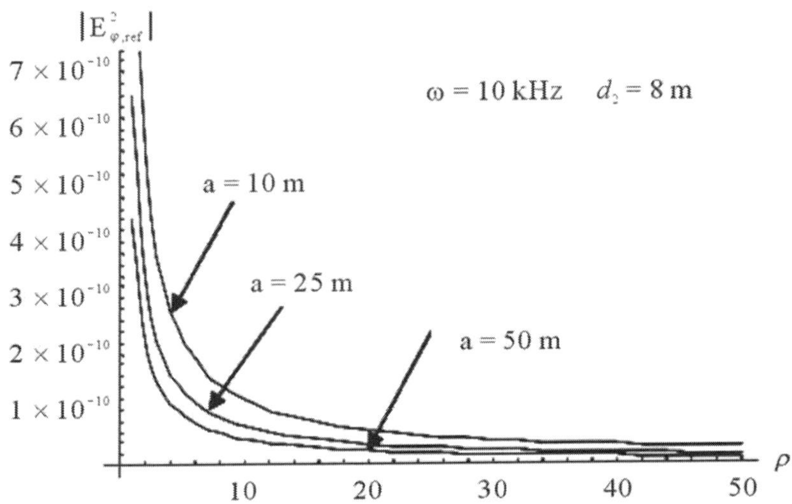

Figure 11. Variation of the absolute value of the reflected electric fields $E^2_{\varphi,ref}$ versus radial distance of the field point for different depths.

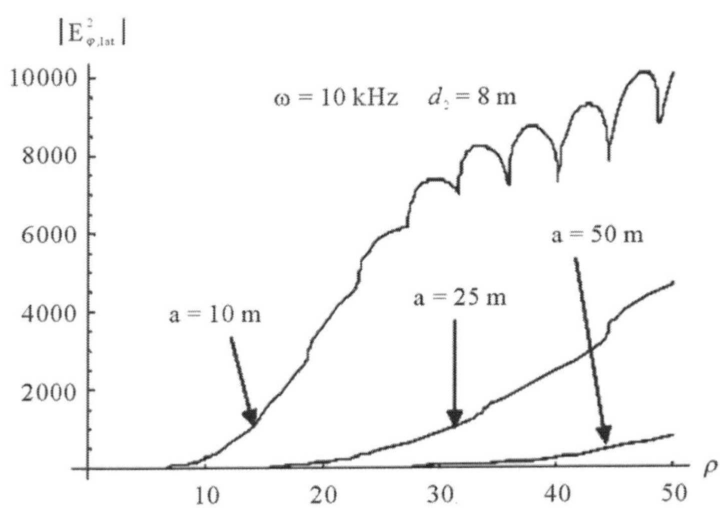

Figure 12. Variation of the absolute value of the lateral electric fields $E^2_{\varphi,lat}$ versus radial distance of the field point for different depths.

6. CONCLUSIONS

In this paper, we summarized our research work regarding the closed-form expressions for the far fields in the sea. The formulas describe the reflected electric field and the lateral waves near the sea surface for different frequencies 10, 20, 30 KHz. The presented results prove that the lateral waves-in the case of uniform surface are very useful in the navigation with very low frequencies as indicate in Figure 8.

For instance in Figures 8-10, we proved that when we fix the radial distance factor $(\rho = 20\,\text{m})$ and the sea depth $(a = 100\ \text{m})$, the lateral waves decreased from 2, 1.5 and 1 as the frequencies decreased from 30, 20 and 10 respectively. Also, from these results we show that the reflected waves can be neglected with respect to the higher values of the lateral waves.

REFERENCES

1. R. W. P. King, "Lateral Electromagnetic Pulses Generated by a Vertical Dipole on a Plane Boundary between Dielectrics," Journal of Electromagnetic Waves and Applications, Vol. 2, No. 3-4, 1988, pp. 225-243.
2. S. T. Bishay, "Communication of Radio Waves in Sea with a Rough Upper Surface," AEÜ Archiv Elektronik Übertrag, Vol. 38, No. 3, 1984, pp. 186-188.
3. L. El S. Rashid, S. T. Bishay and O. M. Abo-Seida, "Electromagnetic Field Produced by a Vertical Magnetic Dipole above a Rough Surface," Proceeding of the International Conference on Electromagnetics in Advanced Applications, 12-15 September 1995, pp. 231-240.
4. O. M. Abo-Seida, S. T. Bishay and K. M. El-Morabie, "Far-Field Radiated from Vertical Magnetic Dipole in Sea with a Rough Upper Surface," IEEE Transactions on Geoscience and Remote Sensing, Vol. 44, No. 8, 2006, pp. 2135-2142.
5. Y. Long, H. Jiang and B. Rembold, "Far-region Electromagnetic Radiation with a Vertical Magnetic Dipole in Sea," IEEE

Transactions on Antennas Propagation, Vol. 49, No. 6, 2001, pp. 992-996.

6. W. C. Chew, "A Quick Way to Approximate a Sommerfeld-Weyl-Type Integral," IEEE Transactions on Antennas Propagation, Vol. 36, No. 11, 1988, pp. 1654-1657.

7. P. R. Bannister, "The Image Theory of Electromagnetic Field of a Horizontal Electric Dipole in the Presence of a Conducting Half-Space," Radio Science, Vol. 17, No. 5, 1982, pp. 1095-1102.

4

CHAPTER

QUASI-PLANAR CHIRAL MATERIALS FOR MICROWAVE FREQUENCIES

Ismael Barba1[1], A.C.L. Cabeceira1[1], A.J. García-Collado2[1], G.J. Molina-Cuberos2[1], J. Margineda2[1] and J. Represa1[1]

[1] *1University of Valladolid, Spain*

1. INTRODUCTION

The growing development in the new communication technologies requests devices to perform new features or to improve the old ones. The trend is to develop new artificial materials reproducing well-known properties already present in other frequency ranges (such as optics) or materials with properties inexistent in the nature. Among the first kind, artificial chiral media, based on the random inclusion of metallic particles with chiral symmetry into a host medium are worth to mention (Fig. 1). Nevertheless, the fabrication techniques up-to-date are quite expensive and produce samples not easy to be tailored and with imperfections, such as intrinsic anisotropy and non-homogeneity (non-uniform density and orientation of inclusions), as well as heavy losses.

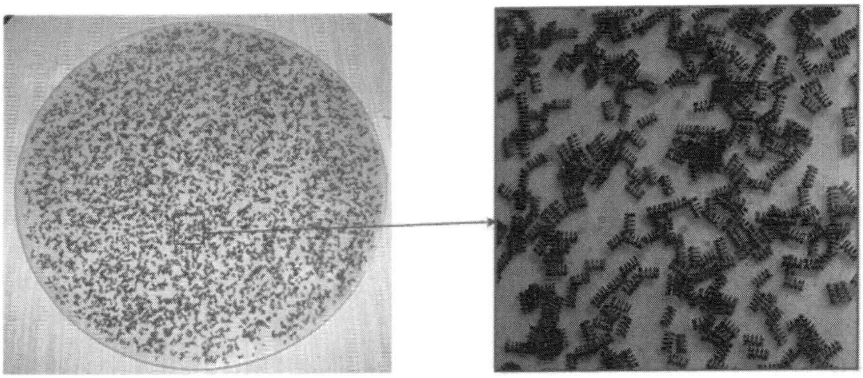

Figure 1. Helix-based artificial chiral material. The sample is a 30 cm diameter disk fabricated by dispersion of six-turn stainless-steel helices in an epoxy resin with a low curing temperature. The helices are 2 mm height and 1.2 mm outer diameter.

During the last years, alternative methods, based on a periodic distribution of planar or quasi-planar chiral particles, have been proposed. This alternative presents the possibility of using conventional printed-circuit fabrication techniques to manufacture the structure. At the same time, the use of via holes provides additional flexibility to select the type of inclusions from helices to cranks or even pseudo-chiral inclusions such as Ω's. As a consequence, the realization of the bulk material, staggering printed circuit plates, gives rise to axial anisotropy.

In this chapter, we are going to present some of the research performed during the last years in the field of chiral materials implementation by means of quasi-planar technologies. Section 2 presents an introduction on chiral materials, as well as (2.2) different approaches in order to implement them: traditional (random) distribution and new, periodic distributions. In the last case, we present different alternative (planar) implementations, finishing with our own (quasi-planar) proposal.

Section 3 shows the two complementary analysis techniques we have employed: numerical analysis, as well as experimental measures, both in free and guided propagation, with a previous fabrication of the samples. Finally, we present (section 4) the results we have obtained (rotation angle of polarization).

2. CHIRAL MATERIALS

Chiral materials are characterized by asymmetric microstructures in such a way that those structures and their mirror images are not superimposable. As a consequence, right- and left-hand circularly polarized waves propagate through the material with different phase velocities and, in case the medium is lossy, absorption rates. Electromagnetic waves in chiral media show the following interesting behavior (Lindell et al., 1994):

1. Optical (electromagnetic) rotatory dispersion (ORD), causing a rotation of polarization;

2. Circular dichroism (CD): due to the different absorption coefficients of a right- and left-handed circularly polarized wave, the nature of field polarization is modified, making linear polarization of a wave to change into elliptical polarization.

These properties have drawn considerable attention to chiral media and may open new potential applications in microwave and millimeter-wave technology: antennas and arrays (Lakhtakia et al., 1988; Viitanen et al., 1998), twist polarizers (Lindell et al., 1992), antireflection coatings (Varadan et al., 1987; Kopyt 2010), etc. It has been also proposed as a way to achieve negative refraction index (Pendry, 2004; Tretyakov et al., 2005). Also, many papers on the analysis of free and guided electromagnetic wave propagation through chiral media have been published, both in time (González-García et al., 1998; Demir et al., 2005; Pereda et al., 2006) and frequency (Xu et al., 1995; Alú et al., 2003; Pitarch et al., 2007; Gómez et al., 2010) domain. For this reason, during the last years, there has been an extensive research on new designs that enhance the above-mentioned properties, as we will see in the next sections of this chapter.

2.1. Constitutive relationships

In contrast to isotropic materials, characterized by their permittivity and permeability, bi-isotropic materials show a cross coupling between electric and magnetic fields, their constitutive relations being:

$$\vec{D} = \varepsilon\vec{E} + \eta\vec{H}$$
$$\vec{B} = \zeta\vec{E} + \mu\vec{H}$$
(1)

where the four scalars ε, μ, η, ζ are function of frequency ω. When the following condition holds:

$$\zeta = -\eta = \frac{j\kappa}{c_0}$$
(2)

c_0 being light speed in vacuum, the medium is said to be "chiral". The parameter

κ

is the "chirality" or "Pasteur" parameter (Lindell et al., 1994). In the frequency domain, this leads to the following constitutive relationships:

$$\vec{D}(\omega) = \varepsilon\vec{E}(\omega) - \frac{j\kappa(\omega)}{c_0}\vec{H}(\omega)$$
$$\vec{B}(\omega) = \mu\vec{H}(\omega) + \frac{j\kappa(\omega)}{c_0}\vec{E}(\omega)$$
(3)

The real part of the chirality parameter is related with the rotation angle of the polarization plane (ORD) in a distance d by means of the following expression:

$$\theta = 2d\omega\frac{\kappa'}{c_0}$$
(4)

Considering electromagnetic field propagation through a homogeneous chiral medium, it is convenient to introduce new field variables, $\vec{E}_{\pm}$ and $\vec{H}_{\pm}$ ("wavefield vectors"), being the following linear combinations of the electric and magnetic fields:

$$\vec{E}_{\pm} = \frac{1}{2}\left(\vec{E} \mp jZ\vec{H}\right), \vec{H}_{\pm} = \frac{1}{2}\left(\vec{H} \pm \frac{j}{Z}\vec{E}\right) \tag{5}$$

where Z is the wave impedance of the medium, $Z = \sqrt{\mu/\varepsilon}$. Actually, the two wavefields $\{\vec{E}_+, \vec{H}_+\}$ and $\{\vec{E}_-, \vec{H}_-\}$ are plane right-circularly and left-circularly polarized waves, respectively. The advantage of introducing these new vectors is that they satisfy the Maxwell equations in an equivalent isotropic medium, so we may use well-known solutions for fields in simple isotropic medium to obtain solutions for wave propagation through chiral media (Lindell et al., 1994). These wavefield vectors will "see" equivalent simple isotropic media with the equivalent parameters:

$$\varepsilon_{\pm} = \varepsilon\left(1 \pm \frac{\kappa}{\sqrt{\varepsilon\mu}}\right), \mu_{\pm} = \mu\left(1 \pm \frac{\kappa}{\sqrt{\varepsilon\mu}}\right) \tag{6}$$

It is clear that, if κ is high enough, one of the wavefield vectors correspond to a backward wave (Tretyakov et al., 2005). That means that, for one of the two possible circularly polarized waves, travelling through a highly chiral material, this one behaves as a left-handed (Veselago) metamaterial.

2.2. Chiral implementations

2.2.1. Random distributions

Traditionally, artificial chiral media at microwave frequencies are fabricated by embedding conducting helices into a host, as shown in Fig 1. The dimensions of these helices determine the bandwidth where the optical activity takes place (Lindman, 1920; Tretyakov et al., 2005). Nevertheless, chirality is a geometrical aspect, therefore helices are not the only possibility, so other type of inclusions, like metal cranks (Molina-Cuberos et al., 2009; Cloete et al., 2001) have been proposed also; an example may be seen in Fig. 2.

Figure 2. Crank-based artificial chiral material. Chiral elements were produced from a 0.4 mm diameter and 12.6 mm length copper wire by bending in three segments by two 90 angles, all with the same handedness. The elements were dispersed in an epoxy resin with a low curing temperature (Molina-Cuberos et al., 2009).

In any case, it is necessary to be careful with the fabrication procedure to assure isotropy and homogeneity. The inclusions must be randomly oriented with no special direction. If the particles are placed in an aligned configuration, the result is a macroscopically bianisotropic material, leading to matrix coefficients for the constitutive parameters (Lindell et al., 1994). Also, a random distribution trends to present local density variations and accidental alignments (see detail in Fig. 1), which causes spatial variations of the constitutive relationships. At the same time, this procedure involves other drawbacks like high cost and difficulty in cutting and molding the material.

In the case of random distribution of cranks, the problems associated to the lack of homogeneity are enhanced. For the same total wire length cranks are bigger than helices, which makes the number density of cranks to be lower than the one using helices and increases the inhomogeneity. Molina-

Cuberos et al. (2009) found fluctuations of the transmitted wave depending on the sample position and orientation with respect to the antenna. Therefore, several measurements and a mean value of the rotation angle were carried out.

2.2.2. Periodical distributions

The problems associated to the lack of homogeneity in chiral media based on random distribution of particles as helices or cranks can be reduced or even eliminated by designing periodical lattices. By an adequate distribution of metallic cranks is possible to build chiral media with homogeneous, isotropic and reciprocal behavior at microwave range (García-Collado et al., 2010), Fig. 3 shows an example of such medium.

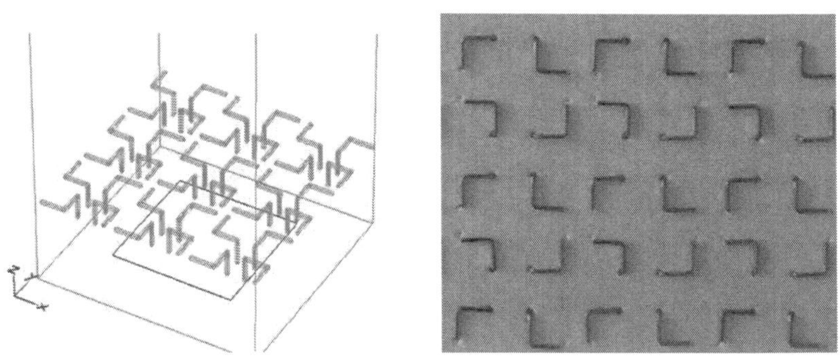

Figure 3. Detailed view of a periodical lattice of cranks with the same handedness produced from 0.68 mm diameter and 15 mm length copper wire by bending in three segments by two 90 degrees angles. One of the segments is introduced perpendicularly into the host medium, polyurethane foam with a relative permittivity close to one. The backside of the medium is completely free of metal. Right: photograph of the lattice (García-Collado et al., 2010). Left: MEFIsTo™ model in which we may see the geometry of the cranks

For these reasons, alternative methods of manufacturing chiral materials have been proposed in recent years: Pendry et al. (Pendry et al.; 2004) proposed a periodical distribution of twisted Swiss-rolls, Kopyt et al. (Kopyt et al.; 2010) a distribution of chiral honeycombs. Nevertheless, most of the alternatives rely on planar and quasi-planar technologies, like Printed Circuit Board (PCB) technology, or even integrated circuit technology, for THz and optical materials. They provide a low-cost technique, which allows a high flexibility in the design of the elementary cell.

Planar technologies make use of two-dimensional elements, in order to obtain media with chiral response. The general concept of chirality, from a geometrical point of view, can be defined in a plane geometry (two dimensions): a structure is considered to be chiral in a plane if it cannot be brought into congruence with its mirror image, unless it is lifted from the plane (Le Guennec, 2000a, 2000b). In this case, it is possible to design a 2D-chiral medium consisting on flat elements possessing no line of symmetry in the plane, and which allows the use of planar technology to manufacture it.

However, electromagnetic activity (electromagnetic rotatory dispersion and circular dichroism) is a phenomenon that takes place in the three dimensional space. Some authors have tried to find electromagnetic activity in thus 2D structures: Papakostas et al. (2003) found a rotation of the polarization plane of a wave incident on a 2D-chiral planar structure like showed in Fig. 4, remarking its apparently nonreciprocal nature: when observed from the back side instead of the front, the sense of the twist is reversed, suggesting then a nonreciprocal polarization rotation similar to that observed in the Faraday effect. That interpretation of this result has opened a discussion on the possibility of a violation of reciprocity and time reversal symmetry (Schwanecke et al., 2003). Kuwata-Gonokami et al. (2005) concluded that such structures are actually chiral in 3D (taking into account air-metal and substrate-metal interfaces), and their electromagnetic activity must arise from this three dimensional nature.

Other groups have achieved three-dimensional chirality by means of multilayered structures of plane elements. The elements may be 2D-chiral (Rogacheva et al., 2006; Plum et al., 2007, 2009) or even non chiral (Zhou et al., 2009): in both cases, the 3D-chirality is obtained by means of a twist between layers (Fig. 5). These structures resulted to give an extremely strong rotation, as well as a negative index of refraction for one of the circularly polarized waves.

Figure 4. Planar chiral structure made by arrays of gammadions arranged in two-dimensional square gratins (Papakostas et al., 2003).

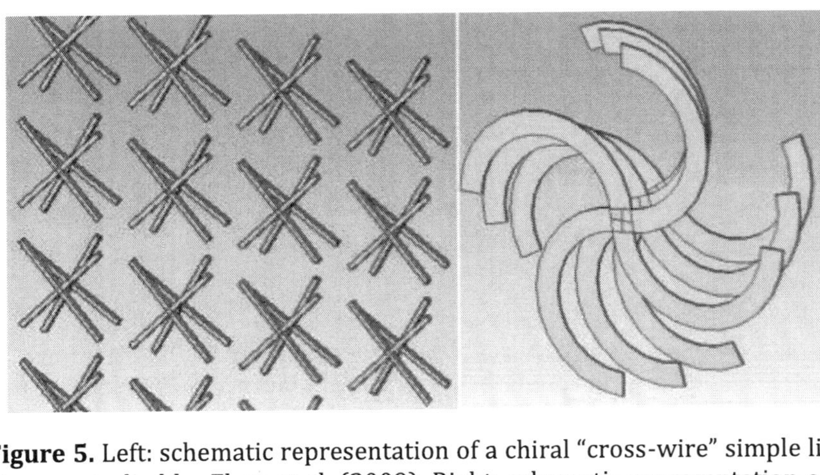

Figure 5. Left: schematic representation of a chiral "cross-wire" simple like the one studied by Zhou et al. (2009). Right: schematic representation of a unit cell of a chiral structure, constructed from planar metal rosettes separated by a dielectric slab (Rogacheva et al., 2006; Plum et al., 2007, 2009)

Finally, it is possible also to construct 3D chiral samples using of quasi-planar technology: in this case, three dimensional PCB technology is employed, involving two-sided boards plus the use of via holes to connect both sides of the board. Such approximation was proposed by Marqués et al. (2007) and also by the authors of this chapter (Molina-Cuberos et al., 2009; Barba et al., 2009).

Figure 6. Photograph of a structure similar to the shown in Fig. 3, but manufactured by means of Printed Circuit technology.

In our research, we have designed different chiral distributions of "molecules" and implemented them by different means: one (Fig. 3) is made by using metal cranks introduced into a polyurethane tablet; the second one (Fig. 6) is, as mentioned, made using PCB technology. We have designed and analyzed different distributions; some of them have been implemented and their behavior measured experimentally, while other ones have been modeled using numerical techniques. More details may be read in the following sections.

3. ANALYSIS

We have worked, first, with the numerical analysis of the designed materials, which allows the study of their electromagnetic behavior at high frequency, previous to the effective construction of the same ones. We have used two different commercially available software in time domain:

- MEFiSTo™, based on TLM method.
- CST Studio Suite™ 2009, based on the finite integration technique (FIT).

Both methods are complete tools to solve electromagnetic problems in 3D, allowing the graphic visualization of the electromagnetic field propagation and its interaction with materials and boundaries during the simulation. The principal advantage of simulating in the time domain is that it most closely resembles the real world. In our case, it allows to obtain a very broadband data with a single simulation run with much less memory requirements than required in frequency-domain methods.

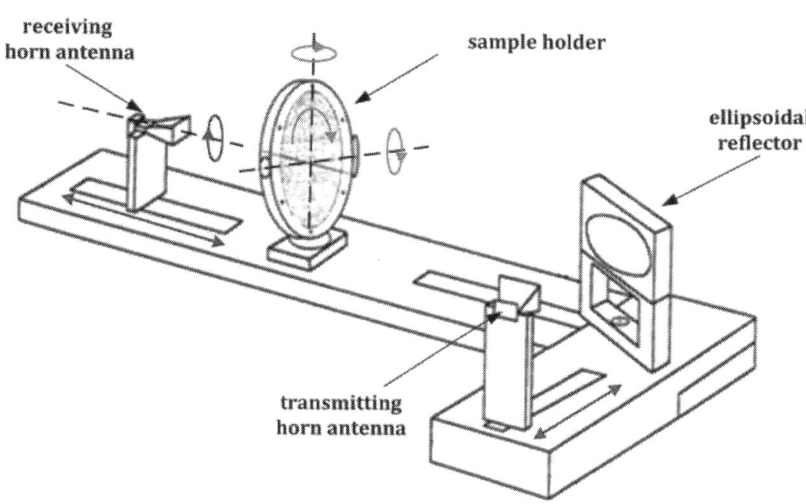

Figure 7. Schematic diagram of the free space setup for the experimental determination of the rotation angle and the three constitutive parameters of isotropic chiral material in the X-Band (not to scale).

The experimental set-up used is based on a previous one for permittivity and permeability measurements at X-band (8.2 - 12.4 GHz) (Muñoz et al., 1998), and adapted to measure electromagnetic activity (Molina-Cuberos et al., 2009; García-Collado et al., 2010). Fig. 7 shows a diagram of the experimental set-up, where the incident wave is linearly polarized in the vertical direction.

The transmitting and receiving antennas are 10-dB-gain rectangular horns. An incident beam is focused by an ellipsoidal concave mirror (30 cm x 26 cm), which produces a roughly circular focal area of about 6 cm in diameter, which is lower than sample size, so that diffraction problems are avoided with relatively small samples. The transmitting antenna is placed at one of the mirror foci (35 cm) and the sample at the other one. The sample holder is midway between the mirror and the receiving antenna and is able to rotate around the two axes perpendicular to the direction of propagation. The receiving antenna, located at 35 cm from the sample, can rotate about the longitudinal axis, which allows the measurement of the scattering parameters (S parameters) corresponding to any polar transmission. The interested reader is referred to Muñoz et al. (1998), Gómez et al. (2008) and García-Collado et al., (2010) for a detailed description of the measurement setup and technique. Here, we briefly present the measurement process:

Figure 8. Waveguide setup for the experimental determination of the rotation angle produced by a chiral material in X-band. A cylindrical sample is located in the circular waveguide and fed through port 1. The transmitted wave is measured in port 2 by a rotating connection, which allows determining the component of the electrical field parallel to the TE_{10} mode of the rectangular wave.

First a two-port "through-reflect-line"' (TRL) calibration is performed at the two waveguide terminals of the network analyzer, PNA-L N5230A, where the antennas are connected. Then, a time domain (TD) transform is used to filter out mismatches from the antennas, edge diffraction effects, and unwanted multiple antenna-mirror-sample reflections or reflections from other parts of the system by means of the ``gating'' TD option of the network analyzer. The rotation angle of the transmitted polarization ellipse is defined as the difference between the polarization direction of the incident wave and the direction of the major axis of the transmitted elliptically polarized wave. Rotation can be determined by looking for the minimum value of the transmitted wave or by measuring the transmission coefficient for co- and cross-polarization, S_{21CO} and S_{21CR} (Balanis, 1989):

$$OA = \left[\frac{1}{2}\left(S_{21CO}^2 + S_{21CR}^2 + \left(S_{21CO}^4 + S_{21CR}^4 + 2S_{21CO}^2 S_{21CR}^2 \cos(2\phi)\right)^{1/2}\right)\right]^{1/2}$$

$$OB = \left[\frac{1}{2}\left(S_{21CO}^2 + S_{21CR}^2 - \left(S_{21CO}^4 + S_{21CR}^4 + 2S_{21CO}^2 S_{21CR}^2 \cos(2\phi)\right)^{1/2}\right)\right]^{1/2}$$

$$\tau = \frac{\pi}{2} - \frac{1}{2}\tan^{-1}\left(\frac{2S_{21CO}S_{21CR}}{S_{21CO} - S_{21CR}}\cos(2\phi)\right) \tag{7}$$

where OA and OB are the major and minor axes, respectively,

ϕ the phase difference between S_{21CO} and S_{21CR}, and

τ the tilt of the ellipse, relative to the incident wave.

In principle, the precise angle of rotation cannot be determined by this measurement alone, there is an uncertainty of

$2n\pi$

, where n is an integer. To determine the angle uniquely, we make use of measurements far away from the resonance range, where it is expected, and found that the rotation angle goes to zero. Once the scattering coefficients are known, it is also possible to retrieve the constitutive parameters $(\varepsilon, \mu, \kappa)$ of the sample.

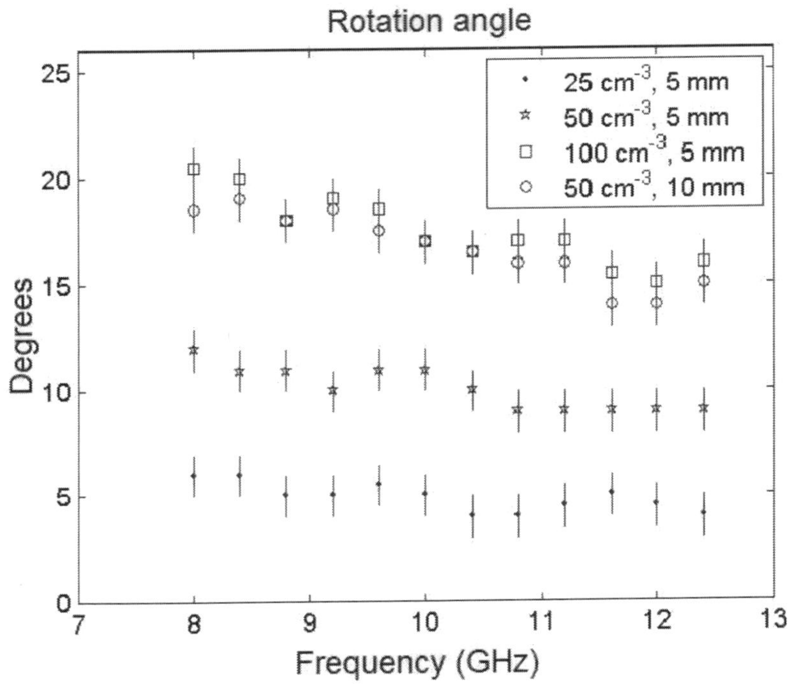

Figure 9. Rotation angle produced by a random distribution of helices (Fig. 1) in a host, for different helix densities (25 cm^{-3}, 50 cm^{-3}, 100 cm^{-3}) and sample thickness (5 mm, 10 mm).

Several analyses of the chiral effects, by making use of waveguide setup, have been also developed; see for example Brewitt-Taylor et al. (1999). In order to test the effect of metallic cranks in waveguide, some samples were initially designed to produce chiral isotropic materials with a resonance frequency at X-band and placed into a section of circular waveguide. Fig. 8 shows the experimental set-up. The sample is excited in a rectangular waveguide and fed to the circular waveguide through a rectangular-circular waveguide transition, the dominant mode in the rectangular waveguide is TE_{10}, and the polarization is perpendicular to the resistive film of the transition which absorbs any cross-polarized field. The dominant mode is TE_{11}, in the empty circular waveguide, and $HE\pm_{11}$ in the chirowaveguide. After the sample, a section of circular waveguide, which can rotate around the longitudinal axis, is connected to a rectangular guide through a transition. The rotation angle of the transmitted wave is obtained by measuring the minimum value of the transmitted wave; we have found that this procedure is more accurate than the determination of the maximum on the transmitted wave.

Figure 10. Schematic illustration of an array of gammadions, chiral in 2D. Each gammadion is assumed to be of copper, and occupies a square of 6x6 mm. There is a separation of 3 mm between each gammadion. The board is 2.5 mm thick.

4. RESULTS

4.1. Random distributions (helices)

Fig. 9 shows the rotation angle produced by a distribution of helices in a host medium for densities ranging from 0 cm^{-3} to 100 cm^{-3}, the error bars showing the uncertainties in the angle determination. Chiral elements are six-turn stainless-steel helices that are 2 mm long and 1.2 mm in outer diameter. The elements were dispersed in an epoxy resin with a low curing temperature. We observe that the rotation angle decreases when the frequency increases, which means that the resonance frequency is below the measurement range. As it can be expected, the rotation angle increases with the number density of inclusions and with the sample width, following a nearly linear relation. Similar behavior has been found in other experiments with helices (Brewitt-Taylor et al., 1999) or cranks (Molina-Cuberos et al., 2005).

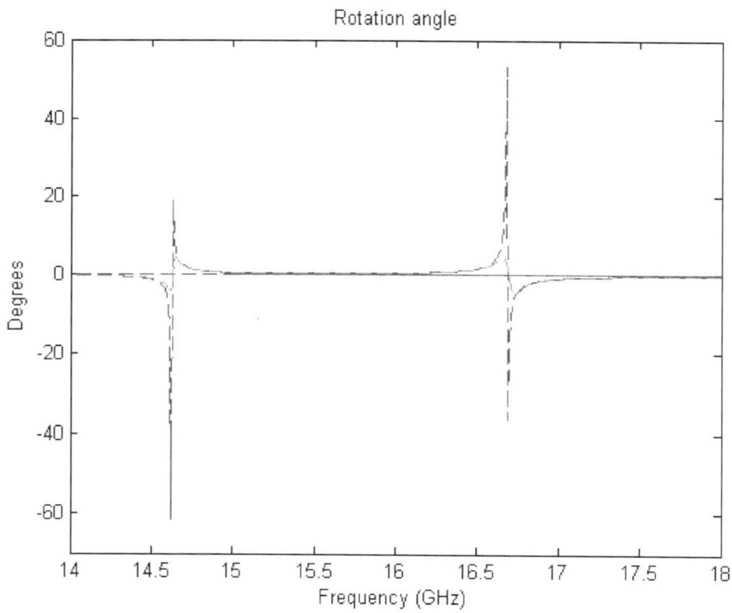

Figure 11. Rotation of the polarization plane for a plane wave normally incident over a planar array of gammadions (Fig. 10), and for different supporting boards: free space (magenta), FR4 (blue), unlossy CER-10 (green) and lossy CER-10 (red). The result is the same in front and back incidence.

4.2. Planar distributions

We have modeled, using CST Studio Suite™ 2009, the rotation of the polarization plane, for a plane wave normally incident over a plane structure, similar, at a different scale, to the one studied by Papakostas et al. (2003). Our structure is also an array of gammadions (Fig. 10) that, in this case, presents resonance in the microwave band. The rotation has been determined assuming different properties of the board that supports the array: first assuming it has the same properties as vacuum, second, a typical material on PC Boards $(\text{FR4},\ \varepsilon_r \approx 4.3)$ and, finally, a high permittivity material, like Taconic CER-10 $(\varepsilon_r \approx 10)$, all present in CST Studio Suite™ 2009 library. The results are shown in Fig. 11. In the first case (vacuum), the structure is symmetrical in a normal axis, so it is not chiral in 3D (the specular image is coincident with the result of a rotation around a longitudinal axis), so there is no electromagnetic activity (no rotation). When taking into account the effect of the board, the structure becomes 3D chiral. In this case, we observe electromagnetic activity, which increases when the properties of the board (permittivity or losses) are higher, i.e., when there is more difference with free space.

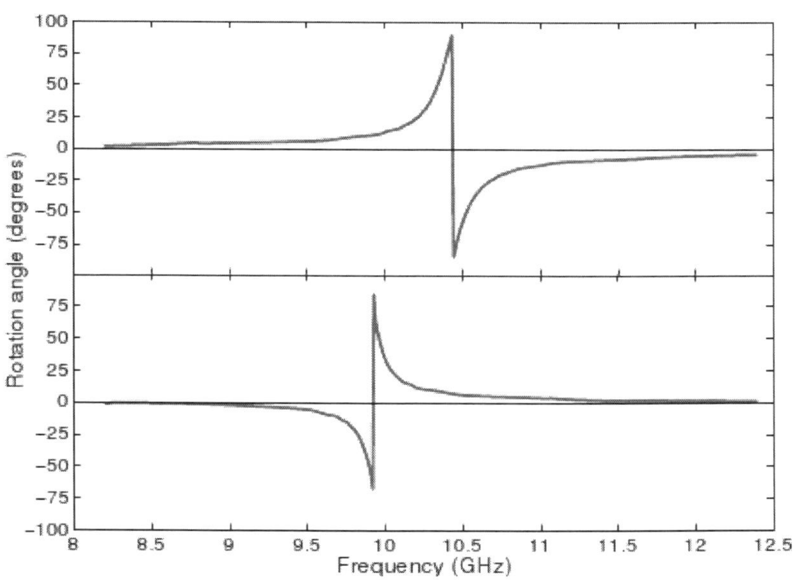

Figure 12. Two examples of the rotation angle produced by a periodical lattice of metallic cranks formed by three equal size segments (5 mm) cranks for left-handed cranks with a separation of 6.9 mm (up) and right-handed cranks with a separation of 9.1 mm (down). [Reprinted from García-Collado et al. (2010) © 2010 IEEE]

4.3. Quasi-planar distributions (cranks)

Fig. 12 shows two examples of the rotation angle produced by periodical lattices of cranks as the one represented in Fig. 3. Both plots correspond to the cranks with the same total length, 15 mm, and different handedness and separation. It can be observed that the sign of the rotation produced by a periodical lattice of cranks depends on the handedness of the elements, as it has been observed in chiral composites formed by randomly oriented elements. In a periodical lattice, the distance of the elements also affects to the characteristic frequencies. In this case, the resonance frequency decreases from 10.4 GHz (up) to 9.8 GHz when the crank separation distance changes from 6.9 mm to 9.1 mm. We do not observe any non-reciprocal effect, i.e. the rotation angle is the same if the wave is incident in the opposite direction.

These results are compared with other ones, obtained by means of time-domain modeling of the same structure, using MeFisTo-3D. In this case, the four cranks of each gammadion are separated 6 mm, while there are 4 mm of distance between two consecutive gammadions. The results are showed in Fig. 13, showing a good agreement between both measures.

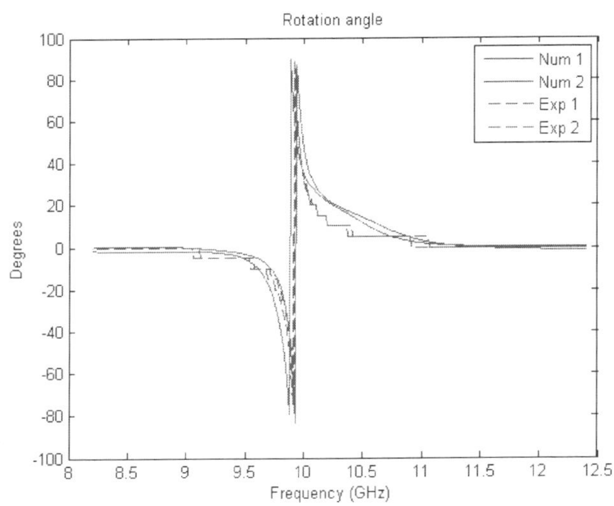

Figure 13. Rotation of the polarization angle for a plane wave normally incident over a quasi-planar periodic array of right-handed cranks as shown in Figs. 3 and 6: numerical (Num) and experimental (Exp) results. 1 and 2 represent the two possible directions of the propagation wave (incident from front and back side, respectively).

Finally, we propose a different distribution of cranks (Fig. 14). In this case, there is a higher concentration of cranks in the same surface, so it is expected to obtain a higher gyrotropy too. That distribution is also geometrically reciprocal.

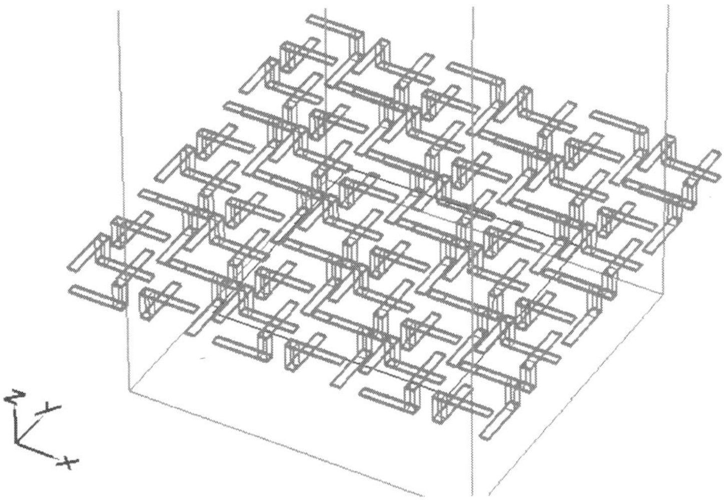

Figure 14. MEFiSTo™ model of a condensed array of cranks. Each crank is composed by two arms, 3mm long, one in each side of the board (1.5 mm of thickness), plus a via connecting both.

The electromagnetic behavior of such distribution has been modeled using MEFiSTo™: we have obtained the rotation of the polarization plane after a normal transmission through that array. The angle of rotation does not depend on the initial polarization of the incident wave (that is, the medium behaves like a biisotropic one, at least in a transversal axis), and it is the same in the two directions of propagation (reciprocal). The result is shown in Fig. 15.

It is worth to mention the couple of discontinuities between -90º and 90º that may be observed in the figure. Such discontinuities are common to most of the distributions we have studied: when we see only one of them (Fig. 12 and Fig. 13) it is caused by the limitations in broadband that suffer our experimental bank. At the same time, other authors (Zhou et al., 2009) find a similar behavior in frequency, being usually assumed to correspond to resonance frequencies. We believe this behavior does not correspond to a real jump in the rotation frequency, but it is a consequence of the measurement procedure, in which the result is normalized between -90º and 90º. If we normalize between 0 y 180º the result in Fig. 15 would be as shown in Fig. 16.

More important: if we study the propagation through several layers of our material, we may draw the rotation angle like in Fig. 17. There, it is demonstrated that the response is lineal (the rotation angle is proportional to the width of the material (number of layers) and, then, the resonance frequency does not depend on the number of layers.

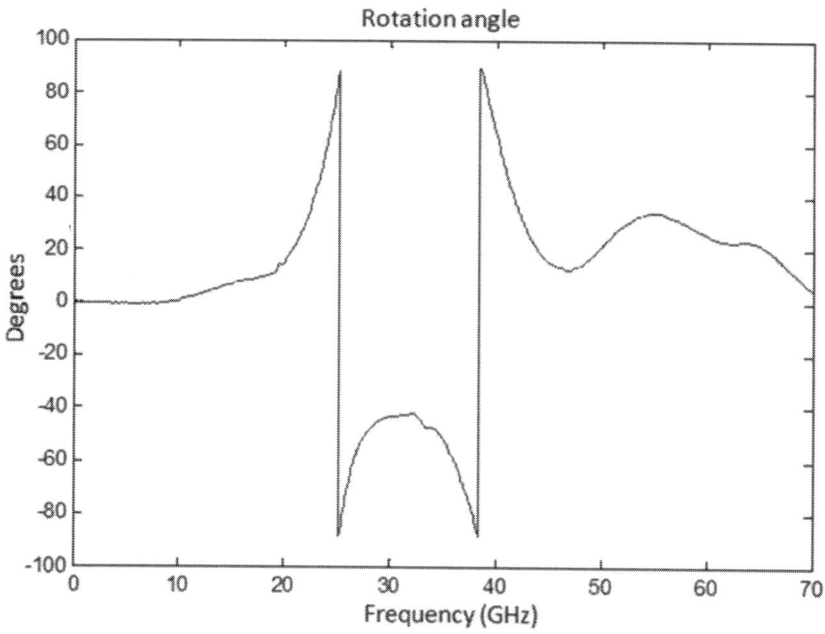

Figure 15. Rotation of the polarization plane for a plane wave normally incident over a condensed array of cranks (Fig. 14), normalizing between -90º and 90º

The chiral material for waveguide experiments was built as described in section 3. However, there are some inherent restrictions in the design due to the limited size of the sample. The radius of the waveguide is similar, in magnitude, to the one of the crank, which strongly limits the number of elements that can be placed on a one-layer distribution, without contact among the elements. Fig. 18 shows two examples produced by four metallic cranks in a foam host medium (left) and eight cranks (right). We have experimentally observed, as it could be deduced by considering symmetry reasons, that other distributions of cranks do not present an isotropic behavior.

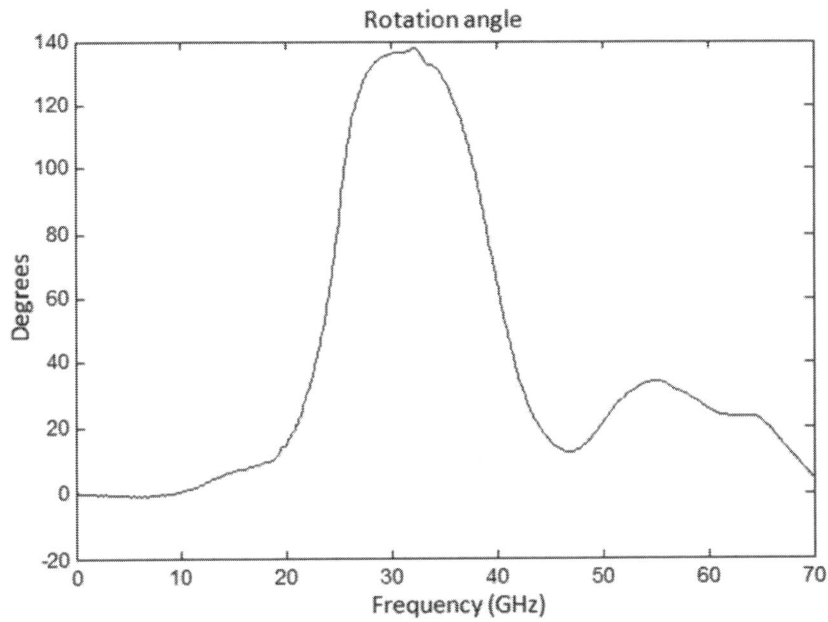

Figure 16. Rotation of the polarization plane for a plane wave normally incident over a condensed array of cranks like represented in Fig. 14, normalizing between 0º and 180º

In order to analyze the response of a single cell, we have measured the rotation angle after a transmission through a group of four cranks, making use of the waveguide setup described in section 3.Fig. 19 shows the rotation angle for cranks formed by equal-size segments, with a total length L ranging from 13.5 mm to 18 mm (Fig. 18). For example, for L = 15 mm, a clear resonance frequency is observed at f_0= 10.08 GHz, the angle is negative below f_0 and positive above f_0. It can be also observed that resonance frequency decreases when the length of the cranks increases, which is in agreement with similar observations found in composites formed by randomly oriented helices (Busse et al., 1999) or cranks (Molina-

Cuberos et al., 2009). The experimental resonance frequencies are 8.24 GHz, 9.04 GHz, 10.1 GHz and 11.7 GHz, very close to a relation $\lambda = 2L$.

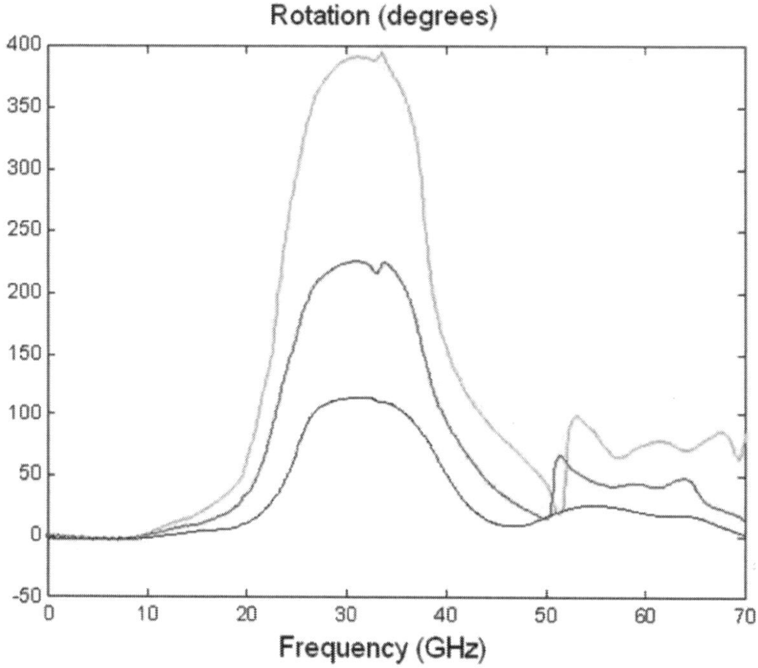

Figure 17. Rotation of the polarization angle for a wave linearly polarized, incident over a condensed distribution of cranks like shown in Fig. 14, for one (blue line), two (red) or three (green) parallel boards. [Reprinted from Barba et al. (2009) © 2009 IEEE]

We have previously checked that the rotation angle does not depend on the relative orientation between cranks and incident wave, i.e. the sample presents an isotropic and homogeneous behavior. This fact does not occur in other configurations with odd number of cranks or with less symmetry properties. In the last case, the observed gyrotropy is a non-chiral effect and other electromagnetic effects, if any, hide the rotation due to chirality. In general we have found isotropic behavior when the sample presents symmetry under 45 degrees rotation, although other rotation symmetries are not ruled out.

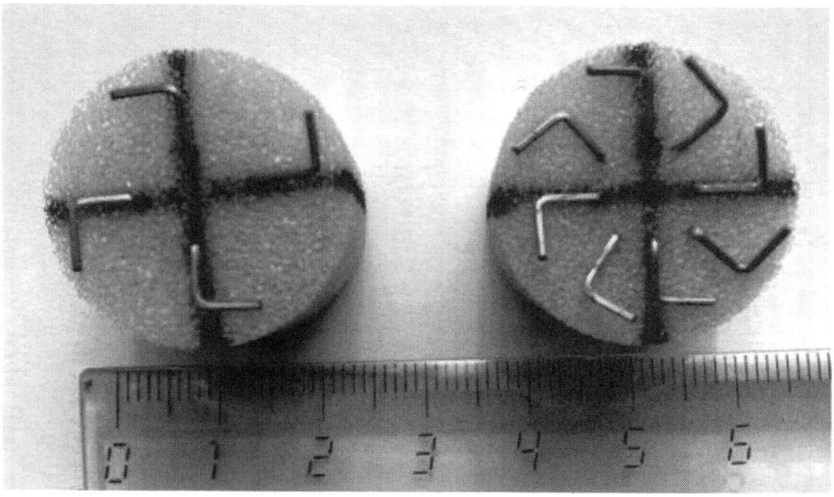

Figure 18. Cylindrical samples used for the experimental determination of chiral effect by using a waveguide setup.

5. CONCLUSION

We have studied different periodical distributions, planar and quasi-planar, which show chiral behavior. We have observed that even when using a planar distribution, its electromagnetic activity comes from its 3D geometry. The rotation will be stronger, then, if we enhance this 3D characteristic. Two possibilities have been studied: some researchers prefer to use multilayered distributions of planar geometries, with a twist between adjacent layers, while we prefer to use two face metallization, with vias connecting both faces of every board: that may present the advantage of obtaining similar electromagnetic activity, combined with thinner structures. The results we have obtained, both using numerical time-domain modeling and experimental measurements seem to support our claim

REFERENCES

1. A. Alú, F. Bilotti, L. Vegni, 2003 Generalized transmission line equations for bianisotropic materials, IEEE Transactions on Microwave Theory and Techniques. 51 11 November 2003), 31343141 . 0018-9480

2. I. Barba, A. C. L. Cabeceira, A. Gómez, J. Represa, 2009 Chiral Media Based on Printed Circuit Board Technology: A Numerical Time-Domain Approach, IEEE Transactions on Magnetics. 45 3 March 2009), 11701173 . 0018-9464

3. A. J. Bahr, K. R. Clausing, 1994 An approximate model for artificial chiral material, IEEE Transactions on Antennas and Propagation. 42 12 December 1994), 15921599 , 0001-8926X.

4. C. A. Balanis, 1989 Advanced Engineering Electromagnetics, John Wiley & Sons, 0-47162-194-3 York, NY, USA.

5. C. R. Brewitt-Taylor, P. G. Lederer, F. C. Smith, S. Haq, 1999 Measurements and prediction of helix-loaded chiral composites, IEEE Transactions on Antennas and Propagation. 47 4 April 1999), 692700 , 0001-8926X.

6. J. H. Cloete, M. Bingle, D. B. Davidson, 2001 The Role of Chirality in Syntehtic Microwave Absorbers, International Journal of Electronics and Communications, 55 4 April 2001), 233239 , 1434-8411

7. E. U. Condon, 1937 Theories of optical rotatory power, Reviews of Modern Physics, 9 (October 1937), 432457 , 0064-6861

8. V. Demir, A. Z. Elsherbeni, Arvas, 2005lsherbeni, A.Z. & Arvas, E (2005) FDTD formulation for dispersive chiral media using the Z transform method, IEEE Transactions on Antennas and Propagation, 53 10 October 2005), 33743384 , 0001-8926X

9. A. J. García-Collado, G. J. Molina-Cuberos, J. Margineda, M. J. Núñez, E. Martín, 2010 Isotropic and homogeneous behavior of chiral media based on periodical inclusions of cranks, IEEE Microwaves and Wireless Components Letters, 20 3 176177 , (March 2010), 1531-1309

10. A. Gómez, A. Lakhtakia, J. Margineda, G. J. Molina-Cuberos, M. J. Núñez, Ipiña. J. A. Saiz, A. Vegas, 2008 Full-Wave hybrid technique

for 3-D isotropic-chiral-material discontinuities in rectangular waveguides: Theory and Experiment, IEEE Transactions on Microwave Theory and Techniques, 56 12 December 2008), 28152824 , 0018-9480

11. A. Gómez, A. Lakhtakia, A. Vegas, M. A. Solano, 2010 Hybrid technique for analyzing metallic waveguides containing isotropic chiral materials, IET Microwaves, Antennas & Propagation, 4 3 March 2010), 305315 , 1751-8725

12. S. González-García, I. Villó-Pérez, R. Gómez-Martín, B. García-Olmedo, 1998 Extension of Berenger's PML for bi-isotropic media, IEEE Microwave and Guided Wave Letters, 8 9 September 1998), 297299 , 1051-8207

13. P. Kopyt, R. Damian, M. Celuch, R. Ciobanu, 2010 Dielectric properties of chiral honeycombs- Modelling and experiment, Composites Science and Technology, 70 7 July 2010), 10801088 , 0266-3538

14. M. Kuwata-Gonokami, N. Saito, Y. Ino, M. Kauranen, K. Jefimovs, T. Vallius, J. Turunen, Y. Svirko, 2005 Giant Optical Activity in Quasi-Two-Dimensional Planar Nanostructures, Physical Review Letters, 95 22 November 2005), 227401 0031-9007

15. A. Lakhtakia, V. V. Varadan, V. K. Varadan, 1988 Radiation by a straight thin-wire antenna embedded in an isotropic chiral medium, IEEE Transactions on Electromagnetic Compatibility, 30 1 February 1988), 8487 , 0018-9375

16. P. Le Guennec, 2000a Two-dimensional theory of chirality. I. Absolute chirality, Journal of Mathematical Physics, 41 9 September 2000), 59545985 , 0022-2488

17. P. Le Guennec, 2000a Two-dimensional theory of chirality. I. Relative chirality and the chirality of complex fields, Journal of Mathematical Physics, 41 9 September 2000), 59866006 , 0022-2488

18. I. V. Lindell, S. A. Tretyakov, M. I. Oksanen, 1992 Conductor-backed Tellegen slab as twist polarizer, Electronic Letters, 28 3 30th January 1992), 281282 , 0013-5194

19. I. V. Lindell, A. H. Sihvola, S. A. Tretyakov, A. J. Viitanen, 1994 Electromagnetic Waves on Chiral and Bi-Isotropic Media, Artech House, 0-89006-684-1 MA, USA.

20. K. F. Lindman, 1920 Uber eine durch ein isotropes system von spiralformigen resonatoren erzeugte rotationspolarisation der elektromagnetischen wellen, Annalen der Physik, 368 23 May 1920), 621644 , 1521-3889

21. R. Marqués, J. Jelinek, F. Mesa, 2007 Negative refraction from balanced quasi-planar chiral inclusions, Microwave and Optical technology Letters, 49 10 October 2007), 26062609 , 0895-2477

22. G. J. Molina-Cuberos, A. J. García-Collado, J. Margineda, M. J. Núñez, E. Martín, 2009 Electromagnetic Activity of Chiral Media Based on Crank Inclusions, IEEE Microwave and Wireless Components Letters. 19 5 May 2009), 278280 . 1531-1309

23. J. Muñoz, M. Rojo, A. Parreño, J. Margineda, 1998 Automatic measurement of permittivity and permeability at microwave frequencies using normal and oblique free-wave incidence with focused beam. IEEE Transactions on Instrumentation and Measurement, 47 4 August 1998), 886892 , 0018-9456

24. A. Papakostas, A. Potts, D. M. Bagnall, S. L. Prosvirnin, H. J. Coles, N. I. Zheludev, 2003 Optical Manifestations of Planar Chirality, Physical Review Letters, 90 10 March 2003), 107404 0031-9007

25. J. B. Pendry, 2004 A Chiral Route to Negative Refraction, Science, 306 5700 November 2004), 13531355 , 0036-8075

26. J. A. Pereda, A. Grande, O. González, A. Vegas, 2006 FDTD Modeling of Chiral Media by Using the Mobius Transformation Technique, IEEE Antennas and Wireless Propagation Techniques, 5 1 December 2006), 327330 , 1536-1225

27. J. Pitarch, J. Catalá-Civera, F. Peñaranda-Foiz, M. A. Solano, 2007 Efficient modal analysis of bianisotropic waveguides by the Coupled Mode Method, IEEE Transactions on Microwave Theory and Techniques, 55 1 January 2007), 0018-9480

28. E. Plum, V. A. Fedotov, A. S. Schwanecke, N. I. Zheludev, Y. Chen, 2007 Giant optical gyrotropy due to electromagnetic coupling, Applied Physics Letters, 90 22 May 2007), 223113 0003-6951

29. E. Plum, J. Zhou, J. Dong, V. A. Fedotov, T. Koschny, C. M. Soukoulis, N. I. Zheludev, 2009 Metamaterial with negative index due to chirality, Physical Review B, 79 3 January 2009), 035407 1098-0121

30. A. V. Rogacheva, V. A. Fedotov, A. S. Schwanecke, N. I. Zheludev, 2006 Giant Gyrotropy due to Electromagnetic-Field Coupling in a Bilayered Chiral Structure, Physical Review Letters, 97 17 October 2006), 177401 0031-9007

31. A. S. Schwanecke, A. Krasavin, D. M. Bagnall, A. Potts, A. V. Zayats, N. I. Zheludev, 2003 Broken Time Reversal of Light Interaction with Planar Chiral Nanostructures, Physical Review Letters, 91 24 December 2003), 247404 0031-9007

32. S. A. Tretyakov, A. H. Sihvola, L. Jylhä, 2005 Backward-wave regime and negative refraction in chiral composites, Photonics and Nanostructures: Fundamentals and Applications, 3 2-3 (December 2005), 107115 , 1569-4410

33. V. K. Varadan, V. V. Varadan, A. Lakhtakia, 1987 On the possibility of designing antireflection coatings using chiral composites, Journal of Wave Matter Interaction, 2 1 January 1987), 7181 , 0887-0586

34. A. J. Viitanen, I. V. Lindell, 1998 Chiral slab polarization transformer for aperture antennas, IEEE Transactions on Antennas and Propagation, 46 9 September 1998), 13951397 , 0001-8926X.

35. Y. Xu, R. G. Bosisio, 21995 An efficient method for study of general bi-anisotropic waveguides, IEEE Transactions on Microwave Theory and Techniques, 43 4 April 1995), 873879 , 0018-9480

36. J. Zhou, J. Dong, W. Wang, T. Koschny, M. Kafesaki, C. M. Soukoulis, 2009 Negative refractive index due to chirality, Physical Review B, 79 12 March 2009), 121104 1098-0121

5

CHAPTER

HIGH FREQUENCY TECHNIQUES: THE PHYSICAL OPTICS APPROXIMATION AND THE MODIFIED EQUIVALENT CURRENT APPROXIMATION (MECA)

Javier Gutiérrez-Meana[1], José Á. Martínez-Lorenzo[1] and Fernando Las-Heras[1]

[1] University of Oviedo, Spain

1. INTRODUCTION

In most of the electromagnetic problems, the number of unknowns to evaluate the scattered fields grows whenever the size of the antenna, device or scenario increases or the working frequency becomes higher. In this context, the rigorous full-wave methods –e.g. Method of Moments (MoM), Fast Multipole Method (FMM) (Engheta et al., 1992), Finite-Difference Time-Domain (FDTD) (Taflove & Umashankar, 1987) or Finite-Difference Frequency-Domain (FDFD) (Rappaport & McCartin, 1991), Finite Element Method (FEM) (Kempel et al., 1998) – can not tackle the analysis of such problems beyond an upper limit determined by the computational requirements in terms of time and memory. High frequency techniques consist in the asymptotic evaluation of the Maxwell's equations. As a consequence, they provide good accuracy when dealing with electrically large geometries meanwhile the computational needs diminish with respect to the aforementioned methods.

Within the high frequency techniques, the Geometrical Optics (GO) and the Physical Optics (PO) approximation are the most extended methods due to the successful results obtained in various fields such as Radar Cross Section (RCS), design of reflector antennas or radioelectric coverage calculation. Since the Physical Optics approximation is detailed in the following section, the Geometrical Optics is briefly summarised.

The main interest in the GO lies in the fact that incident, reflected and transmitted electromagnetic waves are studied based on the conservation of the energy flux along a ray tube between a source and an observation point. Therefore, the Geometrical Optics is usually referred to as Ray Optics. The GO comprises two different methodologies (Rossi & Gabillet, 2002): Ray Tracing (Glassner, 1989) - the starting point is the receiver or observation point and a path to the source is sought analysing the reflections on walls, buildings, mountains - and Ray Launching - multiple rays are launched from the source, so they are independently followed until an observation point or the receiver is reached. One of the common applications of the GO is the evaluation of radio electric coverage or the channel characterization in urban scenarios.

Both the GO and the PO techniques require of an additional method to compute the contribution due to the diffraction phenomenon. The GO can be complemented by means of the Geometrical Theory of Diffraction (GTD) (Keller, 1962) or the Uniform Theory of Diffraction (UTD) (Pathak & Kouyoumjian, 1974). On the other hand, the Physical Theory of Diffraction (PTD) (Ufimtsev, 1962) is applied in joint with the PO formulation.

This chapter focuses on the PO approximation and especially on its extension to dielectric and lossy materials, namely the Modified Equivalent Current Approximation (MECA) method. In section 2, the PO approximation is presented and the formulation is obtained from the equivalence principle. Then the MECA method is introduced, so the equivalent current densities are obtained at the end of section 3.1.3. In order to complete the expressions, the reflection coefficients are then calculated and the determination of the electromagnetic field levels at the observation points based on an analytical solution of the radiation integral is accomplished in section 3.2. A couple of validation examples are studied before continuing to describe the algorithms for solving the visibility problem and, specifically, the Pyramid method in section 4. Afterwards, some of the applications of MECA are listed in "*Application examples*" where an example of radio electric coverage evaluation is also shown. "*Conclusion*" and "*References*" closes the chapter.

2. THE PHYSICAL OPTICS (PO) APPROXIMATION

The Physical Optics (PO) approximation (Harrington, 2001; Balanis, 1989) is a well-known high frequency technique based on the determination of the equivalent current densities induced on the surface of an illuminated perfect electric conductor (PEC) plane. The final expressions can be achieved through the equivalence principle as it will be shown in section 2.1, so once these equivalent or PO current densities are obtained, both electric and magnetic field levels can be calculated from the corresponding radiation integrals.

In order to be in compliance with the constraints introduced by the PO approximation, some aspects have to be evaluated prior to the selection of this technique to tackle any electromagnetic problem:

- The geometry must be made of electrically large obstacles with a smooth variation of their surfaces.
- Given a radiating source, a distinction between lit and non-lit regions must be possible to perform.

The latter implies the need of a complementary algorithm to identify line of sight (LOS) directions from a specific point of view. Some methods will be commented in section 4.

Henceforth, for the sake of simplicity, a time-harmonic variation, $e^{j\omega t}$, of the electromagnetic fields and current densities is assumed. Likewise, the spatial dependence of those variables is not explicitly written.

2.1. PO formulation

The expressions for the electric and magnetic equivalent current densities, J^{PO} and M^{PO} respectively, can be derived from the physical equivalent (Balanis, 1989) in Fig. 1(left), where a scheme consisting of two different media is presented: a PEC on the left and a medium with permittivity ε_1, permeability μ_1 and conductivity σ_1 on the right. Consequently, the boundary *S* between the first and the second media is a PEC plane with the outward normal unit vector $\hat{n}$ pointing to the right.

The incident electric and magnetic fields due to external sources and in absence of any obstacle are E_1^{inc} and H_1^{inc}. On the other hand, the total fields inside the PEC are null $(E_{PEC}^{tot} = H_{PEC}^{tot} = 0)$, while in the second medium

E_1^{inc} and H_1^{inc} are calculated by adding those incident fields to the reflected ones denoted by E_1^{inc} and H_1^{inc}. Therefore the electric and magnetic induced current densities, *J* and *M*, at the boundary S can be obtained from the tangential components of the total fields as:

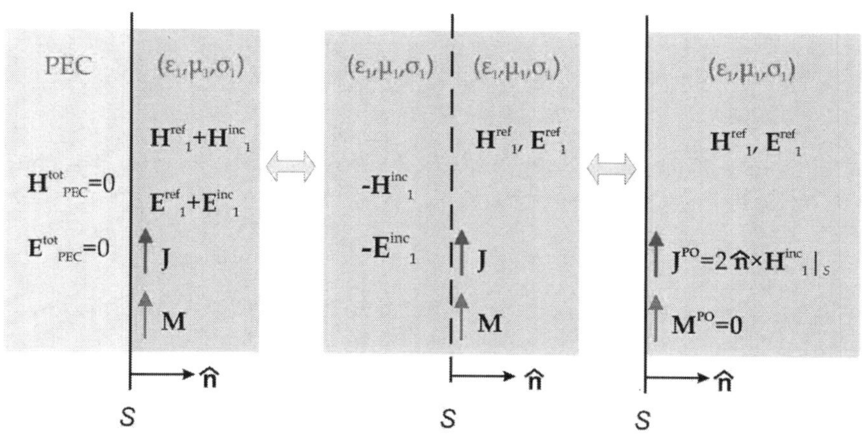

Figure 1. Equivalence principle and Physical Optics approximation.

$$J = \hat{n} \times \left(H_1^{tot} - H_{PEC}^{tot} \right)\Big|_S = \hat{n} \times H_1^{tot}\Big|_S = \hat{n} \times \left(H_1^{inc} + H_1^{ref} \right)\Big|_S$$

$$M = -\hat{n} \times \left(E_1^{tot} - E_{PEC}^{tot} \right)\Big|_S = -\hat{n} \times E_1^{tot}\Big|_S = -\hat{n} \times \left(E_1^{inc} + E_1^{ref} \right)\Big|_S = 0$$

(1)

Fig. 1 (centre) shows the equivalent problem for the non-PEC medium. In order to keep the same boundary conditions, this medium has been extended replacing the PEC part and the total fields are now

$-E_1^{inc}$ and $-H_1^{inc}$ at the left side of S and

E_1^{ref} and H_1^{ref} at its right side.

Considering the characterization of the boundary *S* as a PEC plane in Fig. 1(right) and taking into account the expressions in Eq. (1), the Physical Optics approximation states that $-E_1^{inc}$ and $-H_1^{inc}$ at the boundary *S* are in phase and also have the same amplitude. Thus,

J^{PO} and M^{PO} can be expressed as:

$$J^{PO} = \hat{n} \times \left(H_1^{inc} + H_1^{ref} \right)\Big|_S \approx 2\hat{n} \times H_1^{inc}\Big|_S$$
$$M^{PO} = -\hat{n} \times \left(E_1^{inc} + E_1^{ref} \right)\Big|_S = \mathbf{0} \qquad (2)$$

At this point, an additional consideration to complete the PO formulation is included: when dealing with finite geometries, the PO current density is null in the regions not illuminated by the source:

$$J^{PO} = \begin{cases} 2\hat{n} \times H_1^{inc}\Big|_S & \text{lit region} \\ \mathbf{0} & \text{non-lit region} \end{cases} \qquad (3)$$

This is the reason why the distinction between shadowed and illuminated parts of the scenario is one of the aforementioned constraints to correctly apply the PO approximation.

3. THE MODIFIED EQUIVALENT CURRENT APPROXIMATION (MECA) METHOD: A PO EXTENSION FOR PENETRABLE AND NON-METALLIC OBJECTS

Even though several electromagnetic problems fulfil the restrictions by the PO approximation in terms of electric size or radius of curvature, the characterization of the obstacles as perfect electric conductors reduces the scope of application of this high frequency technique.

Previous works (Rengarajan & Gillespie, 1988; Hodges & Rahmat-Samii, 1993, Sáez de Adana et al., 2004) agree that the extension of the PO approximation for penetrable and non-metallic objects have to account for the reflection coefficients R in order to apply the boundary conditions at the interface between the two media –one of the media uses to be the air. In this context, it is convenient to accomplish a decomposition of the incident fields into their transversal electric (TE) and transversal magnetic (TM) components to independently insert R_{TE} and R_{TM} in the evaluation of the equivalent current densities:

$$E^{inc} = E_{TE}^{inc} + E_{TM}^{inc} = \mathrm{E}_{\mathrm{TE}}^{\mathrm{inc}}\hat{e}_{TE} + \mathrm{E}_{\mathrm{TM}}^{\mathrm{inc}}\hat{e}_{TM}$$

where

$\hat{e}_{TE}$ and $\hat{e}_{TM}$ are the unit vectors in the direction of the TE and TM components of the incident electric field respectively. Likewise, the incident magnetic field can be written in an analogous way.

The Modified Equivalent Current Approximation (MECA) method (Meana et al., 2010) is a new high frequency technique based on the evaluation of a set of equivalent currents to calculate electromagnetic field levels at any observation point. The most important features of this method are:

- MECA deals with electrically large scenarios consisting of dielectric and lossy surfaces due to both electric and magnetic equivalent currents,

J^{MECA} and M^{MECA}, are taken into account.

- MECA reduces to the PO expressions when considering PEC obstacles. Therefore, it can be seen as an extension and an improvement of the classic Physical Optics approximation.
- Reflection coefficients are calculated in a rigorous manner.
- A constant amplitude and linear phase current density distribution on the surface of the facets is adopted to analytically solve the radiation integral.

3.1. MECA formulation

Let us suppose a skew plane wave impinging on the interface

S with an angle of incidence denoted by θ^{inc} and a unit incident vector

$\hat{k}^{inc}$. The two media are now characterised by their respective permittivities

$(\varepsilon_1, \varepsilon_2)$, permeabilities

(μ_1, μ_2)) and conductivities

(σ_1, σ_2). The analysis of the TE and TM components will be presented separately.

3.1.1. TE component

For the incident wave, the coordinate system $\left[\hat{k}^{inc}, \hat{e}_{TM}, \hat{e}_{TE}\right]$ where

$$\hat{e}_{TE} = \frac{\hat{k}^{inc} \times \hat{n}}{\left|\hat{k}^{inc} \times \hat{n}\right|}$$

$$\hat{e}_{TM} = \hat{e}_{TE} \times \hat{k}^{inc} \tag{5}$$

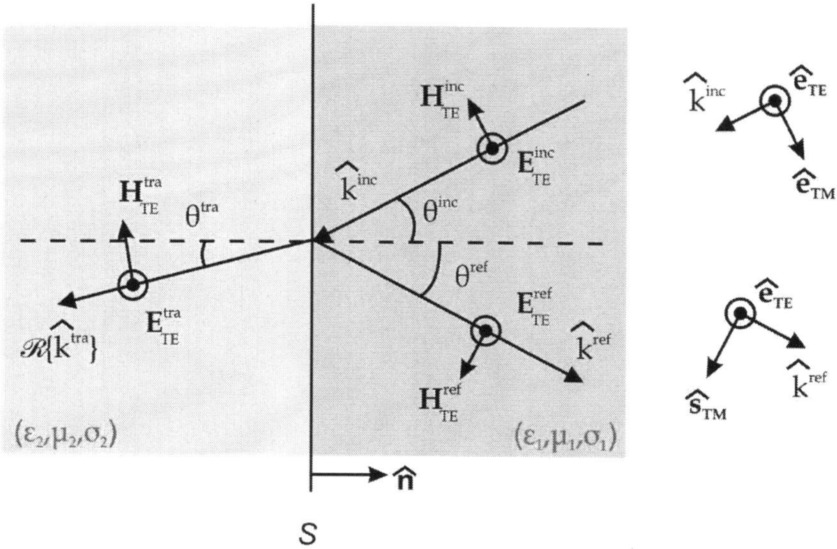

Figure 2. Field decomposition. TE component.

has been chosen (see Fig. 2). Substituting

$\hat{k}^{inc}$ by the unit reflection vector $\hat{k}^{ref}$, the coordinate system $\left[\hat{s}_{TM}, \hat{k}^{ref}, \hat{s}_{TE} = \hat{e}_{TE}\right]$ is defined. Then, the incident electric and magnetic fields are rewritten in terms of these unit vectors:

$$E_{TE}^{inc} = \mathrm{E}_{\mathrm{TE}}^{\mathrm{inc}} \hat{e}_{TE}$$
$$H_{TE}^{inc} = -\mathrm{H}_{\mathrm{TE}}^{\mathrm{inc}} \hat{e}_{TM} = \frac{1}{\eta_1} \mathrm{E}_{\mathrm{TE}}^{\mathrm{inc}} \left(\hat{k}^{inc} \times \hat{e}_{TE} \right) \tag{6}$$

where the intrinsic impedance of the first medium

η_1 is $\eta_i = \sqrt{\dfrac{j\omega\mu_i}{j\omega\varepsilon_i + \sigma_i}}$, i=1,2

.

From the Snell law, the angle of reflection θ^{ref} is equal to θ^{inc}. Therefore,

$\hat{k}^{ref}$ an be expressed as:

$$\hat{k}^{ref} = \hat{k}^{inc} - 2\hat{n}\left(\hat{k}^{inc} \cdot \hat{n}\right) \tag{7}$$

Inserting the TE reflection coefficient R_{TE}, the electric

E_{TE}^{ref} and magnetic H_{TE}^{ref} reflected fields are:

$$\begin{aligned} E_{TE}^{ref} &= \mathrm{R}_{TE}\mathrm{E}_{TE}^{ref}\hat{e}_{TE} \\ H_{TE}^{ref} &== \frac{1}{\eta_1}\mathrm{R}_{TE}\mathrm{E}_{TE}^{inc}\hat{s}_{TM} = \frac{1}{\eta_1}\mathrm{R}_{TE}\mathrm{E}_{TE}^{inc}\left(\hat{k}_{ref} \times \hat{e}_{TE}\right) \end{aligned} \tag{8}$$

Considering that

$$\begin{aligned} E_{TE,TM}^{tot} &= E_{TE,TM}^{inc} + E_{TE,TM}^{ref} \\ H_{TE,TM}^{tot} &= H_{TE,TM}^{inc} + H_{TE,TM}^{ref} \end{aligned} \tag{9}$$

and introducing Eq. (8) in Eq. (9), the total fields for the TE component are:

$$\begin{aligned} E_{TE}^{tot} &= \mathrm{E}_{TE}^{inc}\left(1 + \mathrm{R}_{TE}\right)\hat{e}_{TE} \\ H_{TE}^{tot} &= \frac{1}{\eta_1}\mathrm{E}_{TE}^{inc}\left[\left(\hat{k}^{inc} \times \hat{e}_{TE}\right) + \mathrm{R}_{TE}\left(\hat{k}^{ref} \times \hat{e}_{TE}\right)\right] \end{aligned} \tag{10}$$

3.1.2. TM component

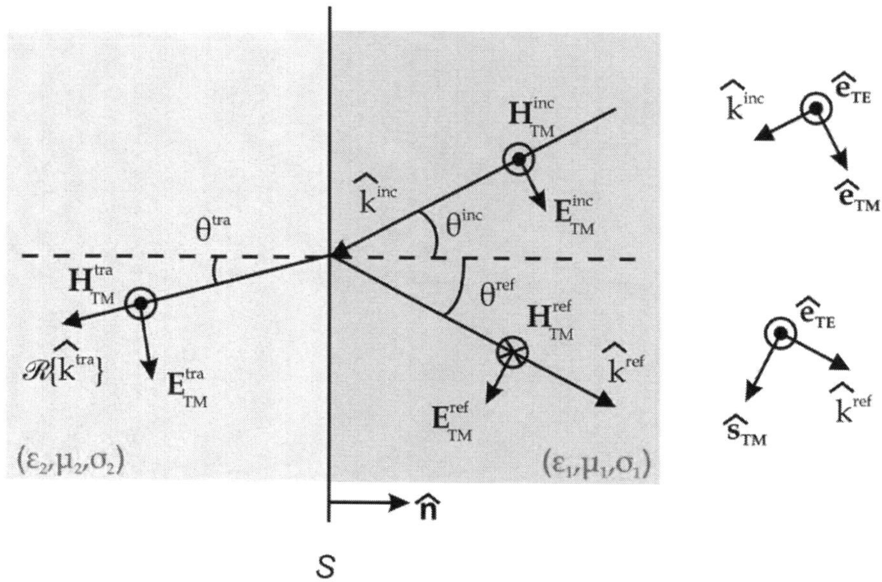

Figure 3. Field decomposition. TM component.

In Fig. 3 the same directions of incidence, reflection and transmission as well as the boundary *S* are again depicted. In addition, the TM components of the fields are plotted. In accordance with the reasoning in the previous section, the expressions for the total fields can be obtained, leading to:

$$E_{TM}^{tot} = \mathrm{E}_{\mathrm{TM}}^{\mathrm{inc}}\left[\hat{e}_{TM} + \mathrm{R}_{\mathrm{TM}}\left(\hat{k}^{ref} \times \hat{e}_{TE}\right)\right]$$

$$H_{TM}^{tot} = \frac{1}{\eta_1}\mathrm{E}_{\mathrm{TM}}^{\mathrm{inc}}\left(1 - \mathrm{R}_{\mathrm{TM}}\right)\hat{e}_{TE} \tag{11}$$

3.1.3. MECA equivalent currents

Adding the contributions of the TE and TM components in Eqs. (10) and (11), the total electric and magnetic fields can be written as a function of the incident electric field, the reflection coefficients, the intrinsic impedance of the first medium, the propagation vectors and the unit vector $\hat{e}_{TE}$:

$$E^{tot} = E^{inc} + \mathrm{R_{TE}E_{TE}^{inc}}\hat{e}_{TE} + \mathrm{R_{TM}E_{TM}^{inc}}\left(\hat{k}^{ref} \times \hat{e}_{TE}\right)$$
$$H^{tot} = \frac{1}{\eta_1}\hat{k}^{inc} \times E^{inc} + \frac{1}{\eta_1}\left[\mathrm{R_{TE}E_{TE}^{inc}}\left(\hat{k}^{ref} \times \hat{e}_{TE}\right) - \mathrm{R_{TM}E_{TM}^{inc}}\hat{e}_{TE}\right]$$

(12)

Going back to the boundary conditions, the MECA equivalent current densities are reckoned from the total fields at S in Eq. (12), being the magnetic ones no longer zero:

$$J^{MECA} = \hat{n} \times H^{tot}\Big|_S =$$
$$= \frac{1}{\eta_1}\left\{\mathrm{E_{TE}^{inc}}\cos\theta^{inc}(1-\mathrm{R_{TE}})\hat{e}_{TE} + \mathrm{E_{TM}^{inc}}(1-\mathrm{R_{TM}})(\hat{n}\times\hat{e}_{TE})\right\}\Big|_S$$
$$M^{MECA} = -\hat{n} \times E^{tot}\Big|_S = \left\{\mathrm{E_{TE}^{inc}}(1+\mathrm{R_{TE}})(\hat{e}_{TE}\times\hat{n}) + \mathrm{E_{TM}^{inc}}\cos\theta^{inc}(1+\mathrm{R_{TM}})\hat{e}_{TE}\right\}\Big|_S \quad (13)$$

As it can be seen and remarked in the previous expressions, the amplitude and phase of the current densities depend on the polarization and also on the angle of incidence θ^{inc} of the incident wave.

3.1.4. TE and TM reflection coefficients

In order to complete the evaluation of the MECA current densities, R_{TE} and R_{TM} have to be determined before moving to the calculation of the electromagnetic fields at the observation points. For this task, the boundary condition for the continuity of the tangential components of the electric and magnetic fields at the boundary S is utilised.

The general dispersion relation applied to the second isotropic medium allows the evaluation of the transmission propagation vector

k^tra

(its real part is depicted in Fig. 2 and Fig. 3) due to

$$k \cdot k = \omega^2 \mu_i \left(\varepsilon_i - \frac{j\sigma_i}{\omega}\right)$$

where

$k = [k_1 \quad k_2 \quad k_3]$ denotes any propagation vector, the subscript *i* refers to the first $(i = 1)$ or second medium (i=2) and$\omega=2\pi f$, with *f* being the working frequency. If$\sigma_i \neq 0$, the right hand side of Eq. (14) is a complex number and the analysis is better performed introducing the real and imaginary parts of the vector *k*, denoted by $\Re\{k\}$ and $\Im\{k\}$ respectively. Particularising for the reflection and transmission propagation vectors:

$$\begin{cases} \left|\Re\left\{k^{ref}\right\}\right|^2 - \left|\Im\left\{k^{ref}\right\}\right|^2 = \omega^2\mu_1\varepsilon_1 \\ 2\Re\left\{k^{ref}\right\}\cdot\Im\left\{k^{ref}\right\} = 0 \\ \\ \left|\Re\left\{k^{tra}\right\}\right|^2 - \left|\Im\left\{k^{tra}\right\}\right|^2 = \omega^2\mu_2\varepsilon_2 \\ 2\Re\left\{k^{tra}\right\}\cdot\Im\left\{k^{tra}\right\} = -\omega\mu_2\sigma_2 \end{cases} \tag{15}$$

The continuity condition at *S* for the first component of the propagation vector means that

$$\Re\left\{k_1^{inc}\right\} = \Re\left\{k_1^{ref}\right\} = \Re\left\{k_1^{tra}\right\} = \frac{2\pi}{\lambda_1}\sin\theta^{inc} \tag{16}$$

where $\lambda 1$ is the wavelength in the first medium and

$$\Im\left\{k_1^{inc}\right\} = \Im\left\{k_1^{ref}\right\} = \Im\left\{k_1^{tra}\right\} = 0 \tag{17}$$

Inserting Eqs. (16) and (17) in Eq. (15), the real and imaginary parts of the third component of the transmission propagation vector are obtained as:

$$\Re\{k_3^{tra}\} = \frac{1}{\sqrt{2}}\left[\Re\{k^{tra2}\} - \left(\frac{2\pi}{\lambda_1}\right)^2 \sin^2\theta^{inc} + \sqrt{\Re\{k^{tra2}\} - \left(\frac{2\pi}{\lambda_1}\right)^2 \sin^2\theta^{inc} + \Im\{k^{tra2}\}}\right]^{1/2}$$

$$\Im\{k_3^{tra}\} = \frac{\Im\{k^{tra2}\}}{2\Re\{k_3^{tra}\}}$$

(18)

Once

$k_3^{tra} = \Re\{k_3^{tra}\} + j\Im\{k_3^{tra}\}$ is determined, R_{TE} is (Staelin et al., 1993)

$$R_{TE} = \frac{\mu_2 \, {}^{2\pi}\!/_{\lambda_1} \cos\theta^{inc} - \mu_1 k_3^{tra}}{\mu_2 \, {}^{2\pi}\!/_{\lambda_1} \cos\theta^{inc} + \mu_1 k_3^{tra}}$$

(19)

and the corresponding R_{TM} is

$$R_{TM} = -\frac{\left(\varepsilon_2 - j\,{}^{\sigma_2}\!/_{\omega}\right){}^{2\pi}\!/_{\lambda_1} \cos\theta^{inc} - \varepsilon_1 k_3^{tra}}{\left(\varepsilon_2 - j\,{}^{\sigma_2}\!/_{\omega}\right){}^{2\pi}\!/_{\lambda_1} \cos\theta^{inc} + \varepsilon_1 k_3^{tra}}$$

(20)

In case of working with non-metallic obstacles, the reflection coefficients become the well-known and simpler Snell's expressions where the calculation of R_{TE} and R_{TM} is straightforward.

When the second medium is a PEC $(\sigma_2 \to \infty)$, the absolute value of the third component of k^{tra} tends to infinite and consequently $R_{TE} = -1$ as well as $R_{TM} = -1$. Substituting these values in Eq. (12), the total magnetic field equals

$2H^{inc}$ at the boundary *S*. Therefore,

J^{MECA} and M^{MECA} in Eq. (13) reduces to

J^{PO} and M^{PO} in Eq. (2).

3.2. Electromagnetic field levels

Once J^{MECA} and M^{MECA} can be calculated with the inclusion of the reflection coefficients in the previous section, the evaluation of the scattered fields due to this set of equivalent current densities on the surface of an obstacle requires solving a radiation integral. For arbitrary distributions and geometries, an analytical solution can not be reached. As a consequence, two assumptions are imposed (Arias et al. 2000; Lorenzo et al., 2005):

1. The incident wave is a plane wave.
2. The obstacle consists of flat triangular facets.

With the objective of distinguishing between the reflected fields at the boundary and the scattered fields at the observation points, the latter are denoted by E^s and H^s . Note that these do not take into account the contribution of the diffraction phenomenon.

3.2.1. Incident plane wave approximation

The position vector $\vec{r}\,'$ of a point on the surface of a flat triangular facet can be written as:

$$\vec{r}\,' = \vec{r}_i + \vec{r}\,'' \tag{21}$$

where $\vec{r}_i$ is the position vector of the barycentre of the i-th facet and $\vec{r}\,''$ is the vector from the barycentre to the source point (see Fig. 4).

The current densities are particularised for the i-th facet, J^{MECAi} and M^{MECAi}, having constant amplitudes, $J_{i0}^{MECA} = J_i^{MECA}\big|_{\vec{r}=\vec{r}_i}$ and $M_{i0}^{MECA} = M_i^{MECA}\big|_{\vec{r}=\vec{r}_i}$ and a linear phase distribution which varies with the unit direction of the Poynting vector as:

$$\hat{p}_i = \frac{\Re\left\{E_i^{inc} \times \left(H_i^{inc}\right)^*\right\}}{\left|\Re\left\{E_i^{inc} \times \left(H_i^{inc}\right)^*\right\}\right|} \tag{22}$$

where E_i^{inc} and H_i^{inc} are the incident electric and magnetic fields at $\vec{r}_i$ respectively, and * denotes the complex conjugate value of the complex vector in brackets. Therefore, in a medium where the working wavelength is λ_1,

$$J_i^{MECA} = J_{i0}^{MECA} \exp\left\{j^{2\pi}/_{\lambda_1}\, \hat{p}_i \cdot \vec{r}\,''\right\}$$
$$M_i^{MECA} = M_{i0}^{MECA} \exp\left\{j^{2\pi}/_{\lambda_1}\, \hat{p}_i \cdot \vec{r}\,''\right\} \tag{23}$$

The reason to introduce the linear phase variation is that this technique allows employing larger facets than when assuming a constant phase distribution.

3.2.2. Application to flat triangular facets

In order to analytically evaluate the radiation integral, the observation points $\vec{r}$ are supposed to be in the far field of the radiating facet – but not necessarily in the far field of the whole scenario. Most of times this constraint is not an additional restriction because applications, such us radar cross section, directly impose huge distances between the object under test and the coordinates where the fields are determined.

Mathematically, the magnitude of the vector between the source point $\vec{r}\,'$ and the observation point $\vec{r}$ is approximated (Balanis, 1989) by the magnitude of the latter $\left|\vec{R}\right| = \left|\vec{r}-\vec{r}\,'\right| \approx \left|\vec{r}\right| \equiv \mathrm{r}$. The phase is taken as $\angle\vec{R} = \angle\vec{r}-\vec{r}\,' \approx \angle \mathrm{r}\text{-}\hat{r}\cdot\vec{r}\,'$, where $\hat{r} = \vec{r}/\left|\vec{r}\right|$

The formulation to compute the scattered fields can be derived from the Maxwell's equations, e.g. using the magnetic vector potential A and the electric potential vector F separately and combining both solutions. The resulting expressions are written in terms of the Green's function in unbounded media G and its gradient $\nabla\mathrm{G}$. The far field approximation allows expressing these functions as:

$$\mathrm{G} = \frac{1}{4\pi}\frac{\exp\left\{-\mathrm{j}2\pi/\lambda_1\left|\vec{R}\right|\right\}}{\left|\vec{R}\right|} \approx \frac{1}{4\pi}\frac{\exp\left\{-\mathrm{j}2\pi/\lambda_1\,\mathrm{r}\right\}}{\mathrm{r}}\exp\left\{\mathrm{j}2\pi/\lambda_1\,\hat{r}\cdot\vec{r}\,'\right\}$$

$$\nabla\mathrm{G} = -\mathrm{j}2\pi/\lambda_1\,\mathrm{G}\,\hat{R}\left(1+\frac{1}{\mathrm{j}2\pi/\lambda_1\,\mathrm{R}}\right) \approx -\mathrm{j}2\pi/\lambda_1\,\mathrm{G}\hat{r} \tag{24}$$

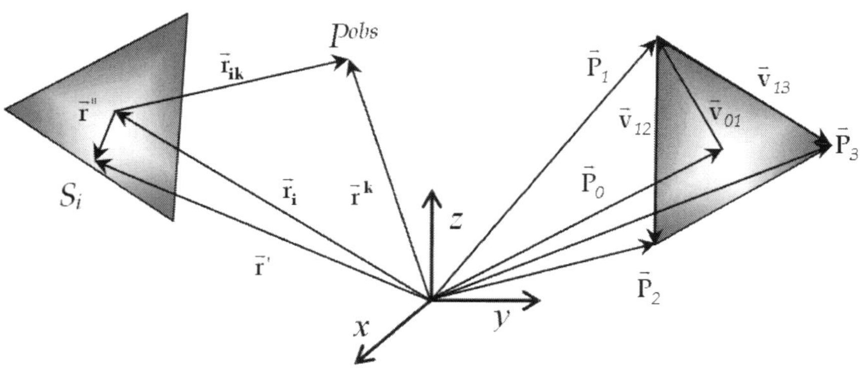

Figure 4. Coordinate system and nomenclature for the triangular facets: on the left, for section 3.2.1 and on the right, for section 3.2.2.

Consequently, the scattered fields are:

$$E^s \approx \frac{j}{2\lambda_1}\frac{\exp\left\{-j\frac{2\pi}{\lambda_1}r\right\}}{r}\int_S\left[\hat{r}\times M^{MECA}-\left(\hat{r}\times\eta_1 J^{MECA}\times\hat{r}\right)\right]\exp\left\{j\frac{2\pi}{\lambda_1}\hat{r}\cdot\vec{r}\,'\right\}\partial s'$$

$$H^s \approx -\frac{j}{2\lambda_1}\frac{\exp\left\{-j\frac{2\pi}{\lambda_1}r\right\}}{r}\int_S\left[\hat{r}\times J^{MECA}-\left(\hat{r}\times\hat{r}\times\frac{1}{\eta_1}M^{MECA}\right)\right]\exp\left\{j\frac{2\pi}{\lambda_1}\hat{r}\cdot\vec{r}\,'\right\}\partial s'$$

(25)

and they can be simplified inserting the auxiliary vectors *Ea* and *Ha*

$$E^s \approx \frac{j}{2\lambda_1}\frac{\exp\left\{-j\frac{2\pi}{\lambda_1}r\right\}}{r}\left[E^a-\eta_1 H^a\times\hat{r}\right]$$

$$H^s \approx -\frac{j}{2\lambda_1}\frac{\exp\left\{-j\frac{2\pi}{\lambda_1}r\right\}}{r}\left[H^a-\frac{1}{\eta_1}\hat{r}\times E^a\right]$$

(26)

With

$$E^a = \int_S \hat{r}\times M^{MECA}\exp\left\{j\frac{2\pi}{\lambda_1}\hat{r}\cdot\vec{r}\,'\right\}\partial s'$$

$$H^a = \int_S \hat{r}\times J^{MECA}\exp\left\{j\frac{2\pi}{\lambda_1}\hat{r}\cdot\vec{r}\,'\right\}\partial s'$$

(27)

Considering that the expressions of *Ea* and *Ha* are quite similar, the following reasoning is only performed for the magnetic fields, so the contribution due to the electric current density J^{MECA} on the i-th facet at the observation point $\vec{r}^{\,k}$ is given by:

$$H_{ik}^{a} = \exp\left\{j 2\pi/_{\lambda_1}\, \hat{r}_{ik} \cdot \vec{r}_i\right\} \int_{S_i} \left(\hat{r}_{ik} \times J_i^{MECA}\right) \exp\left\{j 2\pi/_{\lambda_1}\, \hat{r}_{ik} \cdot \vec{r}\,''\right\} \partial s_i' \tag{28}$$

where S_i refers to the surface of the i-th facet and $\hat{r}_{ik}$ is the unit vector from the barycentre of that facet to the observation point *k*. Substituting J^{MECAi} by its expression in Eq. (23), where the constant amplitude and linear phase variation distributions were inserted and extracting the terms from the integrand, Eq. (28) is rewritten as:

$$H_{ik}^{a} = \exp\left\{j 2\pi/_{\lambda_1}\, \hat{r}_{ik} \cdot \vec{r}_i\right\} \left(\hat{r}_{ik} \times J_{i0}^{MECA}\right) \mathrm{I}(\hat{r}_{ik})$$

$$\mathrm{I}(\hat{r}_{ik}) = \int_{S_i} \exp\left\{j 2\pi/_{\lambda_1}\, (\hat{r} - \hat{p}_i) \cdot \vec{r}\,''\right\} \partial s_i' \tag{29}$$

In case of dealing with flat triangular facets, $\vec{r}\,''$ can be expressed as a function of the vectors $\vec{v}_{mn}$ in Fig. 4:

$$\vec{r}\,'' = \vec{v}_{01} + \mathrm{f}\vec{v}_{12} + \mathrm{g}\vec{v}_{13} \tag{30}$$

With $\vec{v}_{mn} = \vec{P}_n - \vec{P}_m$, being $\vec{P}_l$ the position vector of the vertices l=1,2,3 of the facet and $\vec{P}_0$, the position vector of the barycentre. F and g are real coefficients. Taking into account this transformation, the integral in Eq. (29) is converted into:

$$\mathrm{I}(\hat{r}) = 2\mathrm{A} \exp\left\{-j\frac{a+b}{3}\right\} \int_0^1 \int_0^{1-f} \exp\{j(af + bg)\}\, \partial g\, \partial f \tag{31}$$

where A is the area of the facet and the values of a and b are defined as:

$$a = \frac{2\pi}{\lambda_1} \vec{v}_{12} \cdot (\hat{r} - \hat{p}_i)$$

$$b = \frac{2\pi}{\lambda_1} \vec{v}_{13} \cdot (\hat{r} - \hat{p}_i) \tag{32}$$

The solution to the Eq. (31) is obtained by parts, but in order to overcome some singularities, five different cases are detailed (Arias et al. 2000) in Table 1.

Table 1. Solution to the integral in Eq. (31). Different cases are presented to overcome the singularities in the original expression.

a	b	$I(\hat{r})$
Any	Any	$2A\exp\left\{-j\frac{a+b}{3}\right\}\left[\frac{a\exp\{jb\}-b\exp\{ja\}+b-a}{(a\text{-}b)ab}\right]$
0	Any	$2A\exp\left\{-j\frac{b}{3}\right\}\left[\frac{1+jb-\exp\{jb\}}{b^2}\right]$
Any	0	$2A\exp\left\{-j\frac{a}{3}\right\}\left[\frac{1+ja-\exp\{ja\}}{a^2}\right]$
0	0	A
Any	a	$2A\exp\left\{-j\frac{2a}{3}\right\}\left[\frac{\exp\{ja\}(1-ja)-1}{a^2}\right]$

The formulation of the Modified Equivalent Current Approximation method is completed once all the intermediate results that have been described in this section are combined to evaluate the scattered fields at a specific observation point because of one radiating facet. Until this point, the steps to follow are summarised in:

- Decompose the incident electric field into its TE and TM components.
- Obtain the reflection coefficients in Eqs. (19) and (20).
- Calculate the MECA electric and magnetic current densities in Eq. (13).
- Particularise the expressions for the barycentre of every radiating facet.
- Compute the integral $I(\hat{r}_{ik})$ with the help of Table 1.

- Evaluate the auxiliary fields Ea and Ha introducing the expressions of the MECA current densities and the result of the radiation integral.
- Calculate the scattered fields Es and Hs

Afterwards, the superposition theorem is employed to add the contribution due to all the radiating facets in the geometry for the observation point k:

$$H_k^s = \sum_i H_{ik}^s \tag{33}$$

If an evaluation of the line of sight between the radiating sources and the obstacle were necessary because of its shape, this task would be faced at the beginning of the whole.

3.3. Validation schemes

In this section, two canonical examples for different constitutive parameters and geometries are presented: one assures the good behaviour in high frequency and the other consists in the analysis of electrically large surfaces. In the first scenario, the validation is accomplished in terms of the Radar Cross Section (RCS) which is calculated as:

$$\mathrm{RCS} = 4\pi \lim_{r \to \infty} \mathrm{r}^2 \frac{|E|^2}{|E^{inc}|^2} \tag{34}$$

Since the field levels due to the diffraction contribution are not significant in the selected examples, this effect can be neglected in the computation of the RCS. Consequently, the term in the numerator E only takes into account the electric field levels owing to reflections on the surface.

The methods to contrast the results provided by MECA are an analytical solution taken from the references and the full-wave technique the Method of Moments (MoM) (Medgyesi-Mitschang et al., 1994).

3.3.1. High frequency behaviour

As a high frequency technique, MECA has to show an accurate behaviour when the dimensions of the obstacle are large in comparison with the working wavelength. A good example to test this is a sphere, whose monostatic RCS can be obtained theoretically for both PEC (Balanis, 1989) and non-metallic (Van-Bladel, 2007) characterisations. A frequency sweep is performed for a sphere of radius a with relative permittivity and permeability of $\varepsilon_r = 2.2$, $\mu_r = 1.1$ respectively and conductivity of $\sigma = 7\omega\varepsilon_0$, being $\varepsilon_0 = 8.854 \cdot 10^{-12}$, for the lossy case. The region in Fig. 5 for $2\pi\lambda/r>10$ corresponds to the high frequency zone where a great coincidence in the results can be observed.

3.3.2. Electrically large surfaces

In this setup, a resonant horizontal dipole is placed at a fixed position whose height is of *h=1.5* metres above an electrically large flat surface as depicted in Fig. 6. The position of the observation point P_{obs} is determined by two correlated variables: the horizontal distance from the source to P_{obs}. denoted by d and the height of P_{obs}, denoted by H. The angle υ is defined with the purpose of representing the results.

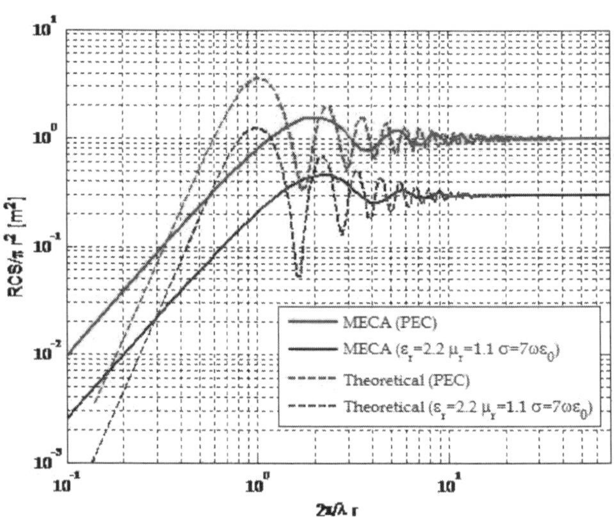

Figure 5. Monostatic RCS of a sphere. PEC and lossy characterization (εr=2.2, μr=1.1, σ=7$\omega\varepsilon 0$).

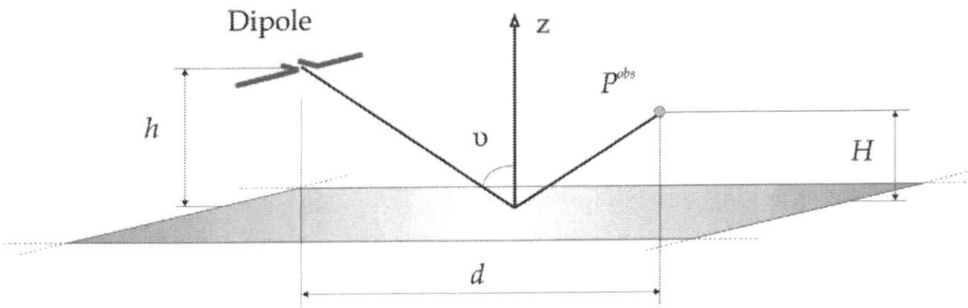

Figure 6. Scheme for testing electrically large surfaces. The dipole is placed horizontally at a height *h*. The position of the observation points varies with the distance from the dipole *d*. For the sake of clarity in the representation of the results, the angle υ has been inserted.

For the working frequency of 1800 MHz and the values of H ranging from 1 to 35 metres, the magnitude of the total electric field $|E^{tot}|$, including the direct illumination from the source, is evaluated. In Fig. 6 MECA results are compared with those by MoM (Medgyesi-Mitschang, 1994) for the following constitutive parameters: $\varepsilon r=3.5$, $\mu r=1$ and $\sigma=0.5S/m$. On the left, $|E^{tot}|$ is plotted as a function of υ, while on the right the variable is the logarithm of the distance d in wavelengths, $d\lambda$, so the small deviation –less than 2dB in the worst case- can be appreciated. Consequently, a high degree of overlapping in the curves of MoM and MECA clearly demonstrates the accuracy of the high frequency technique.

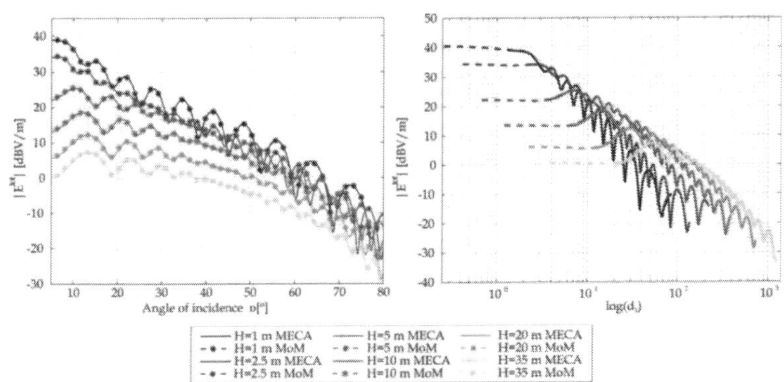

Figure 7. Magnitude of the total electric field for the electrically large surface configuration. On the left, the results have been represented as a function of the angle υ and on the right, as a function of the logarithm of the distance in wavelengths. MECA and MoM show a great coincidence.

4. FAST VISIBILITY ALGORITHM FOR SOLVING THE VISIBILITY PROBLEM

Because MECA (also PO) predicts null equivalent current densities in shadowed regions and non-null equivalent current densities in those regions with line of sight from the source point, a distinction between non-illuminated and illuminated surfaces is mandatory prior to the evaluation of electromagnetic levels at the observation points. This distinction is known as the visibility problem and it is widely found in computer graphics (Dewey, 1988; Foley, 1992; Bittner & Wonka, 2003), e.g. virtual reality or games, but also in the context of electromagnetic problems such as radio electric coverage evaluation or tracking applications in radar.

Within the classic algorithms, the Z-buffer and the Painter's algorithm can be cited. Assuming the screen in the XY plane, both are based on the identification of the z-coordinate of the elements in the geometry to finally project the nearest parts of the objects. Another method is the Binary Space Partitioning (BSP) (Fuch et al. 1980; Gordon & Chen, 1991) which recursively divides the space into front and back semi-spaces, creating a binary tree at the same time. Afterwards, and in accordance with the target direction, the tree is walked and a priority list of the facets can be built.

Most of the techniques to solve the visibility problem are thought to project an image onto a display unit, so the algorithm itself has not better resolution than the pixel size. Therefore, an error is introduced, although almost imperceptible to the human eye. In order to overcome it, solutions where no approximations are made can be implemented. For example, in the Trimming method (Meana et al., 2009) when a piece of surface is partially occluded by other one, only the region in shadow is trimmed and removed from the original geometry. At the end, the remaining surface is the exact part that can be seen from a specific point of view. This type of algorithms is very time consuming but its precision counteracts this disadvantage.

To speed up the computation of the existing techniques, some modifications in the phase of preprocessing can be inserted. One option to carry out this is to split the scenario into parallelepipedic or conic macrodomains. Then the facets are classified in one or other macrodomain depending on their position. As a consequence, a preliminary discrimination is accomplished by blocks, so the facets are discarded faster. In this category, the Angular Z-buffer (AZB) (Cátedra et al., 1998) or the Space Volumetric Partitioning (SVP) (Cátedra & Arriaga, 1999) is included.

Additional acceleration can be achieved when the same algorithms are developed in Graphic Processing Units (GPUs) instead of Central Process Units (CPUs) (Ricks & Kuhlen, 2010). Fortunately, this is not restricted to visibility algorithms but it is suitable for all kind of algorithms consisting of loops, matrix operations, etc. As a consequence, generic GPUs are being extensively utilised to run the codes with smaller costs in comparison with CPU parallelisation. Even more, thanks to the use of libraries, e.g. DirectX or OpenGL (Shreiner, 2004) or the interaction between Matlab® and the graphic cards by means of the free plug-in by Nvidia® or Jacket by AccelerEyes®, the implementation of the routines turns into less complex programming.

In the following section the Pyramid method (Meana et al., 2009) is described due to its suitability to deal with flat triangular facets. Briefly, a right pyramid is built so that its vertex coincides with the source point. The i-th wall is the plane containing the source point and the i-th edge of the facet under test. Thus, a point behind this facet and inside the walls of the pyramid is occluded. This is a fast operation that is accomplished by substituting the coordinates of the observation point in the equation of the plane for all the walls.

4.1. Pyramid method

The scenario consists of N facets which have been sorted by their distance from the source point $\vec{S}$ to their barycentres, so the closest facet, denoted by F^1, is always seen. Additionally, the origin of the coordinate system has been moved to the coordinates of $\vec{S}$. In order to know whether a facet Fj occludes a generic point in the geometry $\vec{P}$, the plane in which Fj is contained is defined:

$$\hat{n}^j \cdot \left(\begin{bmatrix} x & y & z \end{bmatrix} - \vec{P}_0^j \right) = 0 \tag{35}$$

where $\hat{n}^j$ is the unit outward normal vector and $\vec{P}_0^j$ is the barycentre of F^j. Similarly, the three planes containing one of the edges of the Fj and $\vec{s}$ are written as:

$$\frac{\vec{V}_2^j \times \vec{V}_1^j}{\left|\vec{V}_2^j \times \vec{V}_1^j\right|}\cdot \vec{P} = 0\,,\ \frac{\vec{V}_3^j \times \vec{V}_2^j}{\left|\vec{V}_3^j \times \vec{V}_2^j\right|}\cdot \vec{P} = 0\,,\ \frac{\vec{V}_1^j \times \vec{V}_3^j}{\left|\vec{V}_1^j \times \vec{V}_3^j\right|}\cdot \vec{P} = 0 \qquad (36)$$

with $\vec{V}_i^j$ being the position vector of the i-th vertex (i=1,2,3) of F^j. These planes constitute the walls of the pyramid whose base is the facet under test. There are two conditions to conclude that $\vec{P}$ can not be seen from the source point:

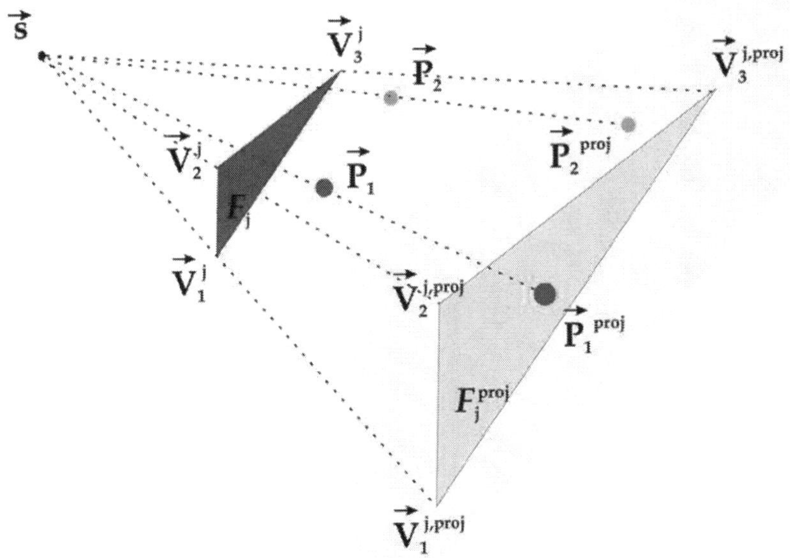

Figure 8. Pyramid method. In the image, a projection plane in light yellow has been selected in order to clearly remark that the facet Fj occludes the

observation point $\vec{P}_1$. On the other hand, $\vec{P}_2$ has line of sight with the source $\vec{s}$.

$\vec{P}$ is behind the plane in Eq. (35)

$$\hat{n}^j \cdot \left(\vec{P} - \vec{P}_0^j\right) < 0 \tag{37}$$

is inside the three walls. An easy way to compute this is:

$$\sum_{i=1}^{3} \text{sign}\left\{\left|\frac{\vec{V}_{i+1}^j \times \vec{V}_i^j}{\left|\vec{V}_{i+1}^j \times \vec{V}_i^j\right|} \cdot \vec{P}\right|\right\} = 3 \tag{38}$$

where $\vec{V}_4^j = \vec{V}_1^j$ and the vertices are supposed to be in clockwise or counterclockwise order.

The algorithm determines the occlusion of points by a triangular facet as depicted in Fig. 8, but the Pyramid method can be applied to determine the existence of line of sight among facets. The process to perform this is based on simplifying the representation of the triangles choosing some important points like the barycentre, vertices, inner points and executing the algorithm for them.

Even though the expressions have been particularised for the case of triangular facets, the implementation can be accomplished for any polygon by including additional walls in the pyramid and substituting the upper limit of the summatory in Eq. (38).

A quite similar algorithm to the Pyramid method is the Cone method (Meana et al., 2009), but a right cone is defined instead of a right pyramid. Its radius is a mean value of the distances from the barycentre of the facet under test to each of its vertices. Analogically, any point behind that facet and inside the cone is occluded. This is a faster algorithm because only one operation comparing the cone angle and the angle between the observation point and the axis of the cone is carried out. On the other hand, the exactness diminishes.

5. APPLICATION EXAMPLES

The scope of application of the Modified Equivalent Current Approximation comprises all the problems which have been analysed by means of the Physical Optics traditionally, but also the extension to dielectric and lossy materials. In the following paragraphs some of these fields where MECA could be employed are summarised.

The RCS computation is one of the most cited topics in the literature: from canonical geometries (e.g. spheres, flat plates and dihedrals) (Griesser & Balanis, 1987; Ross, 1966) for contrasting and validating results to electrically large random surfaces and complex targets (ships or airplanes) (Adana et al., 2000; Uluisik et al., 2008) with a remarkable decrease in the computational time in comparison to full wave techniques. The consideration of absorbing materials allows studying the reduction in the radar signature of aircrafts or missiles.

In order to deal with more realistic scenarios in the RCS computation, the analysis of open-ended cavities including reflections and resonances (Burkholder & Lundin, 2005) is also a line of investigation MECA can face. Likewise, MECA has proven a good accuracy when dealing with electrically large rough surfaces (Meana et al., 2010) that can model the sea surface or the orography of a rural environment. Therefore, the evaluation of the electromagnetic levels for RCS or other applications can consider not only the target itself but also its surroundings.

Another field of application is the design of reflector antennas (Boag & Letrou, 2003; Lorenzo et al., 2005) with or without dielectric radomes because of their electric size. They are usually fed by an array of antennas whose radiation pattern is determined to compute the equivalent current densities and to calculate the fields at distances much larger than the wavelength afterwards.

On the other hand, the Physical Optics approximation and its extensions can also provide some reference values to validate new algorithms or the results by some imaging or shape reconstruction techniques (Saeedfar & Barkeshli, 2006). Due to the fact that the illuminating sources are known a priori, the simulation of a set of constitutive parameters and the shape of the objects can be tackled in a fast manner.

Although the high frequency techniques have lost relevance in the recent years due to the Fast Multipole Method (FMM) –developed by Rohklin in the field of acoustic dispersion (Rohklin, 1985; Rohklin, 1990) and then extended to electromagnetic dispersion (Engheta et al., 1992) – in joint with multilevel schemes and other acceleration methodologies, the advances in the terahertz band will become notable again in the foreseeable future. This means similar applications but working at higher frequencies where the wavelength is smaller than 1 mm and, as a consequence, most of the obstacles and objects is electrically large. Some papers have already proven the validity of the asymptotic approximations at these frequencies.

In addition to the previously enumerated fields of application of the MECA method and in spite of the typical employment of Ray Tracing and Launching in joint with empirical models to deal with the radioelectric coverage evaluation (Papkelis et al., 2007), this problem must be added to that list. The reason is that it fulfils all the requisites detailed through this chapter. In the next section an application example will be shown.

Figure 9. Rural environment for evaluating the radioelectric coverage. The scenario is 17 km long by 10 km width and the 13 base stations have been depicted in red at their real emplacements.

5.1. Radioelectric coverage

Network planning and optimization in rural and urban environments can be studied for different radiocommunication systems (*Global System for Mobile Communications* –GSM –, *Universal Mobile Telecommunications System* –UMTS – or radio and television broadcasting). This example focuses on the radioelectric coverage evaluation of the *General Packet Radio Service* –GPRS – at the working frequency of 1800 MHz, where the scenario consists of a rural terrain of 17 km long by 10 km wide with a uniform soil characterization, $\varepsilon r=3$, $\mu r=1$, $\sigma=0.001 S/m$. Some relevant information about the scenario is that is mainly compound of mountains and valleys where there are not big villages. Thus, regions with no coverage are expected there while, on the other hand, a town on the right and a freeway at the bottom of the map suggest a better coverage in these areas.

The radio electric stations are located at their real emplacements (see Fig. 10) simulating their electromagnetic behaviour based on the available data, such as configuration parameters (e.g. power, gain), provided by the telecommunications company. As a consequence, the radiation patterns were synthesised assuming a dipole array with a different number of elements for every different antenna. Once the illumination from the sources is explicitly determined, the MECA equivalent current densities are computed and the electromagnetic field levels are calculated.

The power density at the observation points situated at 1.5 metres above the terrain is represented in Fig. 10. A threshold has been obtained by taking into account the sensibility of the receivers, so that the minimum in the colorbar has been fixed in a lower value. This means that any region in dark blue can not initiate or maintain a communication with the base station. In order to improve the coverage in specific zones, new base stations could be added and a fast evaluation can be accomplished by computing only the direct illumination due to those stations.

Figure 10. Power density in dBW/m^2 for the rural scenario with MECA. The electromagnetic field levels have been evaluated for a penetrable characterization of the terrain given by $\varepsilon r=3, \mu r=1, \sigma=0.001S/m$.

6. CONCLUSION

In this chapter *"High frequency techniques: the Physical Optics approximation and the Modified Equivalent Current Approximation (MECA)"* the whole process to compute electromagnetic field levels based on the high frequency technique Modified Equivalent Current Approximation has been presented. MECA is an extension of the Physical Optics formulation for penetrable and non-metallic objects based on the equivalence principle, so not only electric but also magnetic equivalent current densities are taken into account for dielectric and lossy materials.

One of the most relevant features of MECA is that the reflection coefficients are calculated from the general dispersion relation for the transversal electric and transversal magnetic components independently. Therefore any conductivity and relative permittivity and permeability are allowed. Then, they are inserted in the corresponding TE/TM analysis which was shown in Fig. 2 and Fig. 3, to be finally combined in order to obtain the MECA equivalent current densities. With the objective of determining the electromagnetic fields at the observation points analytically, the incident wave is supposed to be a plane wave which impinges on the surface and generates a current density distribution with constant amplitude and linear phase variation. Assuming a flat triangular facet, the radiation integral can be solved by parts.

The good behaviour was proven in the validation examples, where the results from the frequency sweep in the high frequency region agreed the theoretical values. Likewise, an excellent overlapping was obtained for different angles of incidence when dealing with a non-PEC electrically large surface.

Because one of the constraints to employ PO and MECA is the determination of line of sight between the source and the observation points, some algorithms to solve the visibility problem were described. The classic methods are complemented by acceleration techniques and then, they are translated into the GPU programming languages. The Pyramid method was explained as an example of fast algorithm which was specifically developed for evaluating the occlusion by flat facets. Undoubtedly, this can be employed in joint with the MECA formulation, but the Pyramid method can also be helpful in other disciplines of engineering.

Throughout the section *"Application examples"*, the way MECA becomes a powerful and efficient method to tackle different scattering problems for electrically large scenarios was satisfactorily demonstrated by means of the example consisting in the evaluation of the radio electric coverage in a rural environment. In addition to this, other fields of application were suggested from the RCS computation to imaging techniques, covering a wide range of electromagnetic problems.

REFERENCES

1. M. Arias, J. O. Rubiños, I. Cuiñas, A. G. Pino, 2000 Electromagnetic Scattering of Reflector Antennas by Fast Physical Optics Algorithms, Recent Res. Devel. Magnetics, 1 1 4363

2. F. S. Adana, P. de Lozano, F. Gisbert, I. Sudupe, J. Pérez, M. F. Cátedra, 2000 Application of the PO to the Computation of the Monostatic RCS of Arbitrary Bodies Modeled by Plane Facets of Dielectric and Magnetic Material, Proceedings of 2000 USCN/URSI National Radio Science Meeting, Salt Lake City, Utah, USA, 2000

3. A. Balanis, 1989 Advanced Engineering Electromagnetics, John Wiley & Sons, 978-0-47162-194-2 New York, USA

4. J. Bittner, P. Wonka, 2003 Visibility in Computer Graphics, Environment and Planning B: Planning and Design, 30 5

5. Boag, C. Letrou, 2003 Fast Radiation Pattern Evaluation for Lens and Reflection Antennas. IEEE Transactions on Antennas and Propagation, 51 5 10631068

6. R. J. Burkholder, T. Lundin, 2005 Forward-Backward Iterative Physical Optics Algorithm for Computing the RCS of Open-Ended Cavities. IEEE Transactions on Antennas and Propagation, 53 2 793799

7. M. F. Cátedra, J. Pérez, Adana. F. Sáez de, O. Gutiérrez, 1998 Efficient Ray-Tracing Techniques for Three Dimensional Analyses of Propagation in Mobile Communications: Application to Picocell and Microcell Scenarios. IEEE Antennas and Propagation Magazine, 40 2 1528

8. M. F. Cátedra, J. P. Arriaga, 1999 Cell Planning for Wireless Communications, Artech House, 978-0-89006-601-0 Boston, USA

9. R. Dewey, 1988 Computer graphics for Engineers, Harpercollins College Div., 978-0-06041-670-6 USA

10. N. Engheta, W. D. Murphy, V. Rokhlin, M. S. Vassiliou, 1992 The Fast Multipole Method (FMM) for Electromagnetic Scattering, IEEE Transactions on Antennas and Propagation, 40 6 634641

11. J. D. Foley, 1992 Computer Graphics: Principles and Practice in C (2nd edition), 0-20184-840-6Wesley, USA

12. H. Fuch, Z. M. Kedem, B. F. Naylor, 1980 On Visible Surface Generation by Priori Tree Structures. ACM SIGGRAPH Computer Graphics, 14 3 124133

13. S. Glassner, 1989 An Introduction to Ray Tracing, Academic Press, 978-0-12286-160-4 San Diego, USA

14. Gordon, S. Chen, 1991 Front-to-Back Display of BSP Trees. IEEE Computer Graphics and Applications, 11 5 7985

15. T. Griesser, C. Balanis, 1987 Backscatter Analysis of Dihedral Corner Reflectors Using Physcal Optics and the Physical Theory of Diffraction, IEEE Antennas and Propagation Magazine, 35 10 11371147

16. R. F. Harrington, 2001 Time-Harmonic Electromagnetic Fields (2nd edition), McGraw Hill, 978-0-47120-806-8 USA

17. R. E. Hodges, Y. Rahmat-Samii, 1993 Evaluation of Dielectric Physical Optics in Electromagnetic Scattering, Proceedings 1993 Antennas and Propagation Society InternationalSymposium, USA, 1993

18. J. B. Keller, 1962 Geometrical Theory of Diffraction. Journal of the Optical Society of America, 52 3 116130

19. L. C. Kempel, A. Chatterjee, J. L. Volakis, 1998 Finite Element Method Electromagnetics (1st edition), IEEE, USA

20. J. A. M. Lorenzo, A. G. Pino, I. Vega, M. Arias, O. Rubiños, 2005 ICARA: Induced-Current Analysis of Reflector Antennas. IEEE Antennas and Propagation Magazine, 47 2 92100

21. J. G. Meana, F. Las-Heras, J. Á. Martínez-Lorenzo, 2009 A Comparison Among Fast Visibility Algorithms Applied to Computational Electromagnetics. Applied Computational Electromagnetics Society Journal, 24 3 268280

22. J. G. Meana, J. Á. Martínez-Lorenzo, F. Las-Heras, C. Rappaport, 2010 Wave Scattering by Dielectric and Lossy Materials Using the Modified Equivalent Current Approximation (MECA). IEEE Transactions on Antennas and Propagation, 58 11 37573761

23. L. N. Medgyesi-Mitschang, J. M. Putnam, M. B. Gedera, 1994 Generalized Method of Moments for Three-Dimensional Penetrable Scatterers, Journal of the Optical Society of America A, 11 4 13831398

24. P. H. Pathak, R. G. Kouyoumjian, 1974 A Uniform Geometrical Theory of Diffraction for an Edge in a Perfectly Conducting Surface. Proceedings of the IEEE, 62 11 14481461

25. G. Papkelis, I. Psarros, I. C. Ouranos, C. G. Moschovitis, K. T. Karakatselos, E. Vagenas, H. T. Anastassiu, P. V. Frangos, 2007 A Radio-Coverage Prediction Model in Wireless Communication Systems Based on Physical Optics and the Physical Theory of Diffraction. IEEE Antennas and Propagation Magazine, 49 2 156165

26. C. M. Rappaport, B. J. Mc Cartin, 1991 FDFD Analysis of Electromagnetic Scattering in Anisotropic Media Using Unconstrained Triangular Meshes. IEEE Transactions on Antennas and Propagation, 39 3 345349

27. S. R. Rengarajan, E. S. Gillespie, 1988 Asymptotic Approximations in Radome Analysis. IEEE Transactions on Antennas and Propagation, 36 3 405414

28. T. Ricks, T. Kuhlen, 2010 Accelerating Radio Wave Propagation Algorithms by Implementation on Graphics Hardware, In: Wave Propagation in Materials for Modern Applications, Andrey Petrin, 103122 , Intech, Vienna, Austria

29. V. Rokhlin, 1985 Rapid Solution of Integral Equations of Classical Potential Theory. Journal of Computational Physics, 60 9 187207

30. V. Rokhlin, 1990 Rapid Solution of Integral Equations of Scattering Theory in Two Dimensions. Journal of Computational Physics, 86 2 414439

31. R. A. Ross, 1966 Radar Cross Section of Rectangular Flat Plates as a Function of Aspect Angle. IEEE Transactions on Antennas and Propagation, 14 8 329335

32. J. P. Rossi, Y. Gabillet, 2002 A Mixed Launching/Tracing Method for Full 3-D UHF Propagation Modeling and Comparison with Wide-Band Measurements. IEEE Transactions on Antennas and Propagation, 50 4 517523

33. Saeedfar, K. Barkeshli, 2006 Shape Reconstruction of Three-Dimensional Conducting Curved Plates Using Physical Optics, NURBS Modeling, and Genetic Algorithm. IEEE Transactions on Antennas and Propagation, 54 9 24972507

34. Adana. F. Sáez de, I. González, O. Gutiérrez, P. Lozano, M. F. Cátedra, 2004 Method Based on Physical Optics for the Computation of the Radar Cross Section Including Diffraction and Double Effects of Metallic and Absorbing Bodies Modeled with Parametric Surfaces. IEEE Transactions on Antennas and Propagation, 52 12 32953303

35. D. Shreiner, 2004 OpenGL Reference Manual: the Official Reference Document to OpenGL,Version 1.4, Addison-Wesley, 978-0-32117-383-6 London, UK

36. D. H. Staelin, A. W. Morgenthaler, J. A. Kong, 1993 Electromagnetic Waves, Prentice Hall, 978-0-13225-871-5 USA

37. Taflove, K. R. Umashankar, 1987 The Finite Difference Time Domain FD-TD Method for Electromagnetic Scattering and Interaction Problems. Journal of Electromagnetic Waves and Applications, 1 4 363387

38. P. Y. Ufimtsev, 1962 Method of Edge Waves in the Physical Theory of Diffraction, Izd-vo Sov. Radio, 1243 (translated by U.S. Air Force Foreign Technology Division, Wright-Patterson AFB, OH)

39. C. Uluisik, G. Cakis, M. Cakis, L. Sevgi, 2008 Radar Cross Section (RCS) Modeling and Simulation, Part 1: A Tutorial Review of Definitions, Strategies, and Canonical Examples. IEEE Antennas and Propagation Magazine, 50 1 115126

40. J. Van -Bladel, 2007 Electromagnetic Fields (2nd edition), IEEE Press,Wiley-Interscience, 978-0-47126-388-3 USA10.1002/047012458X

6

CHAPTER

WAVE PROPAGATION IN NANOCOMPOSITE MATERIALS

Pierre Hillion

Institut Henri Poincaré, Bis Route de Croissy, Le Vésinet, France.

ABSTRACT

Electromagnetic wave propagation is first analyzed in a composite material mde of chiral nano-inclusions embedded in a dielectric, with the help of Maxwell-Garnett formula for permittivity and permeability and its reciprocal for chirality. Then, this composite material appears as an homogeneous isotropic chiral medium which may be described by the Post constitutive relations. We analyze the propagation of an harmonic plane wave in such a medium and we show that two different modes can propagate. We also discuss harmonic plane wave scattering on a semi-infinite chiral composite medium. Then, still in the frame of Maxwell-Garnett theory, the propagation of TE and TM fields is investigated in a periodic material made of nano dots immersed in a dielectric. The periodic fields are solutions of a Mathieu equation and such a material behaves as a diffraction grating.

Keywords: Composite materials, Maxwell-Garnett, Constitutive relations, TE TM Fields

1. INTRODUCTION

Nanotechnology is blossoming with in particular the inclusion of nano-particles (nano-dots) in some specific support [1,2]. Then, to analyze electromagnetic wave propagation such as light or X rays in these composite materials, we need a theory able to calculate average, macroscopic values from their granular microscopic properties. This job is performed by the Maxwell-Garett theory [3-6].

In this work, the support is a dielectric with permittivity e1, permeability m and we consider two situations according that nano-dots are chiral or periodically distributed along a direction of the structure.

In the first case (chiral nano-dots) the permittivity e of the composite material is according to the MaxwellGarett formula

$$[\varepsilon - \varepsilon_1][\varepsilon + 2\varepsilon_1]^{-1} = f\,[\varepsilon_2 - \varepsilon_1][\varepsilon_2 + 2\varepsilon_1]^{-1} \quad (1)$$

f is the filling factor of inclusions (their volume fraction) in the host material, the subscripts 1, 2 corresponding to host and inclusions respectively and we get from (1)

$$\varepsilon = \varepsilon_1\,(1 + 2\alpha f)(1 - \alpha f)^{-1},$$
$$\alpha = (\varepsilon_1 - \varepsilon_2)(\varepsilon_1 + 2\varepsilon_2)^{-1}, \quad (2)$$

Permeability μ is assumed the same for nano-dots and dielectric.

The relation (2) has been generalized [7,8] to chirality x when both inclusions and host materials are chiral. But here, the situation is different since only inclusions have this property and the relation (1) with x, x1, x2 instead of e, e1, e2 has no meaning when x1 = 0. To cope with this difficulty, we introduce a reciprocal Maxwell-Garnett relation obtained by applying to (1) the transformation (x1, x2, f) Þ (fx2, x1, 1/f) which gives

$$[\xi - f\xi_2][\xi + 2f\xi_2]^{-1} = f^{-1}[\xi_1 - f\xi_2][\xi_1 + 2f\xi_2]^{-1} \quad (3)$$

reducing for x1 = 0 to

$$\xi = -2f\xi_2(1 - f)(1 + 2f)^{-1} \quad (4)$$

From now on, we assume f << 1, e1 > 0, e2 < 0, a > 0 and x2 < 0 so that the 0(f 2) approximation of (2) and (4) gives

$$\varepsilon = \varepsilon_1(1 + 3\alpha f) > 0 \tag{5}$$

$$\xi = -2f\xi_2 > 0 \tag{6}$$

So, this composite material made of nano-chiral particles included in a dielectic may be hand-led as an homogeneous chiral medium with permittivity and chirality (5) and (6) and permeabiity μ > 0 assumed to be the same for inclusions and dielectric.

In the second situation (periodically distributed nanorods), the relation (1) is still valid with f changed into a periodic function f (x). Assuming f (x) = f cos(2ax), we write the permittivity e (x) in the following form reducing to (5) to the 0(f 2) order

$$\varepsilon(x) = \varepsilon_1 \exp[3\alpha f \cos(2ax)] \tag{7}$$

Using (5)-(7), we shall analyze harmonic plane wave propagation in both composite materials, chiral and periodic.

2. Harmonic Plane Wave Propagation in a Chiral Composite Medium

We suppose this chiral medium endowed with the Post constitutive relations in which e, x have the expressions (5) and (6) [9,10]

$$\mathbf{D} = \varepsilon\mathbf{E} + i\xi\mathbf{B},\ \mathbf{H} = \mathbf{B}/\mu + i\xi\mathbf{E},\ i = \sqrt{-1} \tag{8}$$

This choice is not arbitrary because the Post constitutive relations, in their general form, are covariant under the proper Lorentz group as Maxwell's equations which guarantees a consistent theory with a simple mathematical formalism, in agreement with the statement that only covariant mathematical expressions have a physical meaning.

Plane wave scattering from a semi-infinite chiral medium was discussed some time ago by Bassiri et al [11], also using the Post constitutive relations, but we proceed differently from these authors working with the Fresnel reflection and transmission amplitudes.

2.1 Refractive Indices

We consider harmonic plane waves with amplitudes E, B, D, H

$$(\underline{\mathbf{E}},\underline{\mathbf{B}},\underline{\mathbf{D}},\underline{\mathbf{H}})(\mathbf{x},t) = (\mathbf{E},\mathbf{B},\mathbf{D},\mathbf{H})\,\psi(\mathbf{x},t) \tag{9}$$

in which

$$\psi(\mathbf{x},t) = \exp[i\omega(t + n\sin\theta\, x/c + n\cos\theta\, z/c)] \tag{10}$$

in which n is a refractive index to be determined.

Substituting (9) into the Maxwell equations

$$\begin{aligned} \nabla\wedge\underline{\mathbf{E}} + 1/c\partial_t\underline{\mathbf{B}} = 0,\ \nabla.\underline{\mathbf{B}} = 0 \\ \nabla\wedge\underline{\mathbf{H}} - 1/c\partial_t\underline{\mathbf{D}} = 0,\ \nabla.\underline{\mathbf{D}} = 0 \end{aligned} \tag{11}$$

and taking into account (10) give the equations for the amplitudes E, B, D, H

$$\begin{aligned} -n\cos\theta\, E_y + B_x = 0, \\ n(\cos\theta E_x - \sin\theta E_z) + B_y = 0, \\ n\sin\theta\, E_y + B_z = 0, \\ n\cos\theta H_y + D_x = 0 \end{aligned} \tag{12}$$

$$\begin{aligned} n(\cos\theta H_x - \sin\theta H_z) - D_y = 0 \\ n\sin\theta\, H_y - D_z = 0 \end{aligned} \quad 13)$$

with the divergence equations

$$\sin\theta B_x + \cos\theta B_z = 0,\ \sin\theta D_x + \cos\theta D_z = 0 \tag{14}$$

We get at once from (8) and (14), the divergence equation satisfied by the electric field sinq

$$\sin\theta\ E_x\ +\ \cos\theta\ E_z\ =\ 0 \tag{15}$$

Substituting (8) into (13) gives

$$n\cos\theta\,(B_y/\mu + i\xi E_y) + \varepsilon E_x + i\xi\, B_x = 0$$

$$n\cos\theta\,(B_x/\mu + i\xi E_x) - n\sin\theta\,(B_z/\mu + i\xi E_z) - \varepsilon E_y - i\xi B_y = 0$$

$$n\sin\theta\,(B_y/\mu + i\xi E_y) - \varepsilon E_z - i\xi B_z = 0 \tag{16}$$

Taking into account (12), these equations become

$$n\cos\theta\, B_y/\mu + \varepsilon E_x + 2i\xi B_x = 0$$

$$n\cos\theta\, B_x/\mu - n\sin\theta\, B_z/\mu - \varepsilon E_y - 2i\xi B_y = 0$$

$$n\sin\theta\, B_y/\mu - \varepsilon E_z - 2i\xi B_z = 0 \tag{17}$$

Then, eliminating B between (12) and (17) gives the homogeneous system of equations in $\sigma = 2n\xi$

$$(n^2/\mu - \varepsilon)E_x - i\sigma\cos\theta\, E_y = 0$$

$$(n^2/\mu - \varepsilon)E_y - i\sigma(\sin\theta\, E_z - \cos\theta\, E_x) = 0$$

$$(n^2/\mu\ -\ \varepsilon)E_z\ +\ i\sigma\sin\theta\ E_y\ =\ 0 \tag{18}$$

This homogeneous system has nontrivial solutions if its determinant is null and a simple cal-culation gives

$$(n^2/\mu\ -\ \varepsilon)[(n^2/\mu\ -\ \varepsilon)^2\ -\ \sigma^2]\ =\ 0 \tag{19}$$

Deleting (n2/m - e) = 0 which would correspond to an a-chiral medium, we get from (11) two modes (n±2/m - e) = ± s in which which s = 2nx so that the refractive index depends not only on permittivity and permeability but also on chirality with the positive expressions

$$n_+ = \xi\mu + (\xi^2\mu^2 + \varepsilon\mu)^{1/2}, \tag{20}$$

$$n_- = -\xi\mu + (\xi^2\mu^2 + \varepsilon\mu)^{1/2} \tag{21}$$

Changing the square root into its opposite gives negative refractive indices.

Consequently, two modes with respectively the refractive indices n+, n- can propagate in the metachiral slab, they are independent as long as the medium is infinite, otherwise they become coupled at boundaries. The amplitudes of the field components in these two modes have now to be determined.

2.2 Electromagnetic Fields

1) We first suppose $n_+^2/\mu - \varepsilon = \sigma$ and $n_+ = \xi\mu + (\xi^2\mu^2 + \varepsilon\mu)^{1/2}$ with e and μ > 0: fields and parameters are characterized by superscripts or subscripts + respectively.

Then, we get at once from (18) and (12) in terms of E^+_y

$$\begin{aligned} &E^+_x = i\cos\theta_+ E^+_y,\ E^+_z = -\,i\sin\theta_+ E^+_y,\\ &B^+_x = n_+ \cos\theta_+ E^+_y,\\ &B^+_y = -i\ n_+ E^+_y,\ B^+_z = -\ n_+ \sin_+ E^+_y \end{aligned} \tag{22}$$

and substituting (22) into (8)

$$\begin{aligned} &D^+_x = i\cos\theta_+ \lambda_+ E^+_y,\\ &D^+_y = \lambda_+ E^+_y,\\ &D^+_z = -i\sin\theta_+ \lambda_+ E^+_y \end{aligned} \tag{23}$$

$$H^+_x = \cos\theta_+ \ \nu_+ E^+_y,$$
$$H^+_y = -i \ \nu_+ E^+_y,$$
$$H^+_z = -\sin\theta_+ \ \nu_+ E^+_y$$

(24)

in which

$$\lambda_+ = \varepsilon + \xi n_+, \ \nu_+ = n_+/\mu - \ \xi = (\xi^2 + \varepsilon/\mu)^{1/2}$$

(25)

2) For $n_-^2/\mu - \varepsilon = -\sigma$ and $n_- = -\xi\mu + (\xi^2\mu^2 + \varepsilon\mu)^{1/2}$, we get at once with now super-scripts and subscripts -:

$$E^-_x = -i\cos\theta_- E_y, \ E^-_z = i\sin\theta_- E^-_y,$$
$$B^-_x = n_- \cos\theta_- E^-_y, \ B^-_y = i\, n_- E^-_y,$$
$$B^-_z = - n_- \sin\theta_- E^-_y$$

(26)

and substituting (26) into (8)

$$D^-_x = -i\cos\theta_- \ \lambda_- E^-_y, \ D^-_y = \lambda_- E^-_y,$$
$$D^-_z = i\sin\theta_- \ \lambda_- E^-_y$$

(27)

$$H^-_x = \cos\theta_- \ \nu_- E^-_y, \ H^-_y = i \ \nu_- E^-_y,$$
$$H^-_z = - \sin\theta_- \ \nu_- E^-_y$$

(28)

With

$$\lambda_- = \varepsilon - \xi n_-, \ \nu_- = n_-/\mu + \ \xi = (\xi^2 + \varepsilon/\mu)^{1/2} = \nu_+ \quad (29)$$

Then, according to (9) and (10), the electromagnetic field of the plus and minus modes, each depending on an arbitrary amplitude E+y , E-y, is

$$(\underline{\mathbf{E}}^{\pm}, \underline{\mathbf{B}}^{\pm}, \underline{\mathbf{D}}^{\pm}, \underline{\mathbf{H}}^{\pm})\,(\mathbf{x}, t) = (\mathbf{E}^{\pm}, \mathbf{B}^{\pm}, \mathbf{D}^{\pm}, \mathbf{H}^{\pm})\ \psi_{\pm}(\mathbf{x}, t) \quad (30)$$

with the amplitudes given by (22)-(24) and (26)-(28) and the phase functions

$$\psi_{\pm}(\mathbf{x},t) = \exp[i\omega(t + n_{\pm} \sin\theta_{\pm} x/c + n_{\pm} \cos\theta_{\pm} z/c)] \tag{31}$$

2.3 Plane Wave Scattering from a Semi-Infinite Chiral Composite Medium

We suppose that the chiral composite material fulfills the half space z < 0 on which impinges from z > 0 on the interface z = 0 an harmonic plane wave characterized by the phase factor y(qi)

$$\psi(\theta_i) = \exp[-i\omega n_0(x\sin\theta_i + z\cos\theta_i)] \tag{32}$$

n0 is the refractive index in z > 0 and the components of the incident electromagnetic field are [12] with two amplitudes Mi, Ni:

$$E^i_x = -\cos\theta_i M_i \psi(\theta_i),\ E^i_y = N_i \psi(\theta_i),\ E^i_z = \sin\theta_i M_i \psi(\theta_i)$$
$$H^i_x = -n_0\cos\theta_i N_i \psi(\theta_i),\ H^i_y = -n_0 M_i \psi(\theta_i),$$
$$H^i_z = n_0\sin\theta_i N_i \psi(\theta_i) \tag{33}$$

The reflected field in the half-space z > 0 has a similar expression with (Mi, Ni, qi) changed into (Mr, Nr, qr) while the refracted field in z < 0 is supplied by (30).

According to (31) and (32), also valid for the reflected wave, the continuity of the phase at z = 0 implies the Descartes-Snell relations

$$n_0 \sin\theta_i = n_0 \sin\theta_r = n_+ \sin\theta_+ = n_- \sin\theta_- \tag{34}$$

The continuity of the components $E_{x,y}$, $H_{x,y}$, at z = 0 supplies four boundary conditions to de-termine in terms of Mi, Ni the amplitudes Mr, Nr of the reflected field and those $E^+{}_y$, $E^-{}_y$ of the refracted field.

According to (22), (26) and (33) and taking into account (34), we get for the Ex,y components

$$\cos\theta_i(M_r - M_i) = i\cos\theta_+ E^+_y - i\cos\theta_- E^-_y$$
$$N_r + N_i = E^+_y + E^-_y \quad (35)$$

while for $H_{x,y}$, according to (24), (28) and (33), we have since $\nu_- = \nu_+ (= \nu)$

$$n_0 \cos\theta_i(N_r - N_i) = \nu(\cos\theta_+ E^+_y + i\cos\theta_- E^-_y)$$
$$n_0(M_r + M_i) = \nu(E^+_y - E^-_y) \quad (36)$$

To make calculations easier, we introduce the notations

$$M_r + M_i = M,\ N_r + N_i = N,$$
$$M_r - M_i = M',\ N_r - N_i = N' \quad (37)$$

And

$$a = n_0/\nu\ (\cos\theta_+ + \cos\theta_-)^{-1} \quad (38)$$

Then, we get at once from (36)

$$E^+_y = a(\cos\theta_i N' + \cos\theta_- M)$$
$$E^-_y = a(\cos\theta_i N' - \cos\theta_+ M) \quad (39)$$

and, substituting (39) into (35) gives

$$\cos\theta_i M' = a_{11} N' + a_{12} M$$
$$N = a_{21} N' + a_{22} M \quad (40)$$

in which

$$a_{11} = ia\cos\theta_i(\cos\theta_+ + \cos\theta_-),\ a_{12} = 2ia\cos\theta_+\cos\theta_-$$

$$a_{21} = a\cos\theta_i,\ a_{22} = a(\cos\theta_- + \cos\theta_+) \qquad (41)$$

Taking into account (37) the system (40) becomes

$$(\cos\theta_i - a_{12})M_r + a_{11} N_r = (\cos\theta_i + a_{12}) M_i - a_{11} N_i$$

$$a_{22}M_r - (1 - a_{21}) N_r = -a_{22}M_i + (1 + a_{21})N_i \qquad (42)$$

from which we easily get the amplitudes Mr, Nr of the reflected field and consequently M', N' according to (37) to obtain finally the amplitudes $E_{y\pm}$ of the refracted field from (39).

One has a simple result for a normal incidence qi = qr = q± = 0 since the Equations (35) and (36) reduce to

$$M_r - M_i = i(E^+_y - E^-_y),\ N_r + N_i = E^+_y + E^-_y$$

$$-2n_0 N_i = \nu(E^+_y + E^-_y),\ n_0(M_r + M_i) = \nu(E^+_y - E^-_y) \qquad (43)$$

with the solution

$$M_r = (\nu + in_0)(\nu - in_0)^{-1} M_i,\ N_r = -(1 + 2n_0/\nu)N_i \qquad (44)$$

$$E^+_y = n_0(\nu - in_0)M_i - n_0/\nu\ N_i,$$

$$E^-_y = -n_0(\nu - in_0)M_i - n_0/\nu\ N_i \qquad (45)$$

Remark 1. If the angles q+, q- obtained from (34) are real, the plus and minus modes propa-gate in the chiral medium. If they are both purely imaginary, we get from (34)

$$\cos(\theta_\pm) = -i[(n_0/\nu_\pm)^2 \sin^2\theta_i - 1]^{1/2} \qquad (46)$$

the negative sign in front of the square root in (46) corresponds to the physical situation: refracted waves are evanescent and, incident waves undergo a total reflection, with as consequence for beams of plane waves a GoösHanken lateral shift and a Imbert-Fedorov transverse shift [13]. Of course with a single angle pure imaginary, only one mode propagates, the other mode giving rise to an evanescent wave.

Remark 2. At the expense of more intricacy, the present formalism may be generalized to wave propagation in a chiral slab located between z = 0 and z = - d. Then, two more fields exist respectively reflected at z = -d inside the slab and refracted outside in the z < -d region, supplying four supplementary amplitudes matched by the boundary conditions at z = -d. But, instead of a 4 ´ 4 system of equations to get the amplitudes of the electromagnetic field, we have to deal with a 8 ´ 8 system more difficult to solve.

3. HARMONIC PLANE WAVE PROPAGATION IN A TWO DIMENSIONAL NANO-PERIODIC MEDIUM

With B = μH, D = e (x)E, and exp(-iwt) implicit, the Maxwell equations are for E(x,z), H(x,z)

$$\partial_z E_y - i\omega\mu/c\, H_x = 0,\ \partial_z H_y + i\omega\, \varepsilon(x)/c\, E_x = 0$$
$$\partial_z E_x - \partial_x E_z + i\omega\mu/c\, H_y = 0,$$
$$\partial_z H_x - \partial_x H_z - i\omega\, \varepsilon(x)/c\, E_y = 0$$
$$\partial_x E_y + i\omega\mu/c\, H_z = 0,\ \partial_x H_y - i\omega\varepsilon(x)/c\, E_z = 0 \quad (47)$$

with the divergence equations

$$[\varepsilon' + \varepsilon\partial_x]E_x + \varepsilon\partial_z E_z(x,z) = 0,\ \partial_x H_x + \partial_z H_z = 0 \quad (48)$$

giving rise to TE (E_y, H_x, H_z) and TM (H_y, E_x, E_z) waves.

3.1 TE Wave Propagation

Assuming f << 1, we work with the Maxwell-Garnett 0(f2) approximation of (7)

$$\varepsilon(x) = \varepsilon_1 + \eta f \cos(2ax), \quad \eta = 3\alpha\varepsilon_1 \tag{49}$$

The component Ey satisfies the Helmholtz equation in which $\Delta = \partial_x^2 + \partial_z^2$

$$[\Delta + \omega^2 \mu \varepsilon(x)/c^2] E_y(x,z) = 0 \tag{50}$$

We look for the solutions of this equation in the form, A being an arbitrary amplitude

$$E_y(x,z) = A \exp(ik_z z)\ \psi(x) \tag{51}$$

Substituting (51) into (50) and taking into account (49), gives the differential equation satisfied by $\psi(x)$

$$[\partial_x^2 + k_0^2 + f k_e^2 \cos(2ax)]\ \psi(x) = 0 \tag{52}$$

in which

$$k_0^2 = \omega^2 \mu \varepsilon_1 / c^2 - k_z^2,\ k_e^2 = \omega^2 \mu \eta / c^2 \tag{53}$$

Using the variable z = k1 x , Equation (52) becomes a Mathieu equation [14,15]

$$[\partial_\zeta^2 + \chi^2 + f \cos(2a\zeta/k_e)]\,\psi(\zeta) = 0,\ \chi^2 = k_0^2/k_e^2 \tag{54}$$

with solutions in the form [14,15,16] where v has to be determined

y (z) = ∑m=-¥¥ cm exp([i(v + 2m) az/ke)] (55)

Substituting (55) into (54) gives the following recurrence relation [15] for the coefficients c_m

$$c_m + \gamma_m(\varpi)\ (c_{m-1} + c_{m+1}) = 0 \tag{56}$$

with

$$\gamma_m(\varpi) = -f/2\,[(2m + \varpi)^2 - \chi^2] \tag{57}$$

Now, the main difficulty [14,15] is to get v in terms of f and c, but f being small, the infinite determinant of the system (56) supplies v to the 0(f 3) order [15]

$$\cos(\varpi\pi) = \cos(\chi\pi) + \pi f^2\,[4\chi^2(1-\chi^2)^{1/2}]^{-1} \sin(\chi\pi) \tag{58}$$

Once v known, the cm coefficients may be obtained by numerical methods based on the recurrence relations (36) or on some variant of it. It is shown [15] how for moderate values of c and f, these relations can be transformed into convergent continued fractions Rm(n) = cm/cm-1, Lm(n) = cm/cm+1.

So, according to (51) and (55), $E_y(x,z) = E_y(x + \pi/a, z)$ and

$$E_y(x,z) = A\,\exp(ik_z z) \sum_{m=-\infty}^{\infty} c_m \exp[i(\varpi + 2m)ax)] \quad 0 \le x < \pi/a \tag{59}$$

and taking into account the Maxwell Equation (47), the other two components H_x, H_z of the T_E field are obtained from $\partial_z E_y$ and $\partial_x E_y$ respectively. Writing (59)

$$E_y(x,z) = A \sum_{m=-\infty}^{\infty} c_m \exp(ik_z z + ik_m x),\quad k_m = (\varpi + 2m)a,\ 0 \le x < \pi/a \tag{60}$$

$E_y(x,z)$ appears as a periodic beam of plane waves propagating in the directions defined by the wave vectors with components (kz, km), their amplitude being weighted by the coefficients cm.

3.2 TM Wave Propagation

For TM waves (H_y, E_x, E_z), we start with the expression (7) of e (x). Then, according to the Maxwell Equation (47) the component Hy satisfied the equation

$$[\Delta + \omega^2 \mu\varepsilon(x)/c^2 - \{\varepsilon'(x)/\varepsilon(x)\}\partial_x]\, H_y(x,z) = 0 \tag{61}$$

We look for the solutions of (61) in the form

$$H_y(x,z) \;=\; A\, \exp(ik_z z)\; \psi(x) \tag{62}$$

$$\psi(x) = u(x)\; \phi(x),\; \phi(x) = \exp[f_1/2\, \cos(2ax)],$$
$$f_1 \;=\; f\,\eta \tag{63}$$

A simple calculation gives the first and second derivative of y (x)

$$\psi'(x) = [u'/u - af_1\, \sin(2ax)]\,\psi(x)$$
$$\psi''(x) = [u''/u - 2a\, u'/u\, f_1\, \sin(2ax)$$
$$-\, 2a^2\, f_1\, \cos(2ax) \;+\; a^2 f_1^{\,2}\, \sin^2(2ax)]\,\psi(x) \tag{64}$$

and since e'/e = -2a f1 sin(2ax), we get to the 0(f12) order

$$\psi'' - \varepsilon'/\varepsilon\; \psi' = [u'/u - 2a^2 f_1\, \cos(2ax)]\;\; \psi(x) + 0(f_1^{\,2}) \tag{65}$$

so that

$$[\psi'' - \varepsilon'/\varepsilon\; \psi']H_y\,(x,z) = [u'/u - 2a^2 f_1\, \cos(2ax)]H_y(x,z) \tag{66}$$

Then, according to (62) and (66), we get from (61), the differential equation satisfied by $u(x)$

$$[\partial_x^2 + \omega^2 \mu\varepsilon(x)/c^2 - k_z^{\,2} - 2a^2 f_1\, \cos(2ax)]u(x) = 0 \tag{67}$$

which becomes with the Maxwell-Garnett approximation (49) of $\varepsilon(x)$

$$[\partial_x^2 + k_0^2 + f k_h^2 \cos(2ax)]u(x) = 0 \qquad (68)$$

with k02 given by (53) while

$$k_h^2 = \omega^2 \mu \varepsilon_1 / c^2 - 2a^2 \eta \qquad (69)$$

The comparison of (52) and (68) shows that, to the 0(f 2) order, one has just to change ke into kh to go from TE to TM waves so that all the calculations of Subsection 3.1 can be repeated mutatis mutandis.

3.3 TE Wave Scattering in a Semi Infinite Nano-Periodic Material

The granular material, made of nano dots immersed in a dielectric, lies in the z < 0 half-space and we suppose that a TE harmonic plane wave (Eyi, Hxi, Hzi) impinges from the upper half-space z > 0 with refractive index n and permeability m on the z = 0 interface.

The components Eyi, Eyr of the incident and reflected waves are Eyi(x,z) = Ai exp[iwn/c (x sinqi + z cosqi)]

$$E_y^i(x,z) = A_i \exp[i\omega\nu/c\ (x \sin\theta_i + z \cos\theta_i)]$$
$$E_y^r(x,z) = A_r \exp[i\omega\nu/c\ (x \sin\theta_i - z \cos\theta_i)] \qquad (70)$$

and according to the Maxwell Equation (47), the components H_x^i, H_x^r involved in the boundary conditions are

$$H_x^i(x,z) = \nu/\mu \cos\theta_i\, E_y^i(x,z),$$
$$H_x^r(x,z) = -\nu/\mu\ \cos\theta_i\ E_y^r(x,z) \qquad (71)$$

Now, the refracted periodic field in z < 0 has the form (59)

$$H_x^i(x,z) = \nu/\mu \cos\theta_i \, E_y^i(x,z),$$
$$H_x^r(x,z) = -\nu/\mu \cos\theta_i \, E_y^r(x,z) \qquad (72)$$

and, still using (47)

$$H_x^t(x,z) = \gamma E_y^t(x,z), \; \gamma = ck_z/\omega\mu \qquad (73)$$

the boundary conditions impose the continuity on z = 0 of Ey and Hx, that is, according to (70)-(73)

$$(A_i + A_r)\exp(i\omega\nu/c\, x \sin\theta_i)$$
$$= A_t \sum\nolimits_{m=-\infty}^{\infty} c_m \exp[i(\varpi + 2m)ax],$$
$$0 \le x < \pi/a \qquad (74)$$

$$\nu/\mu \cos\theta_i (A_i - A_r)\exp(i\omega\nu/c\, x \sin\theta_i)$$
$$= \gamma A_t \sum\nolimits_{m=-\infty}^{\infty} c_m \exp[i(\varpi + 2m)ax],$$
$$0 \le x < \pi/a \qquad (75)$$

Let us write (74)

$$A_i + A_r = A_t \Omega(\beta\pi),$$
$$\Omega(\beta\pi) = \sum\nolimits_{m=-\infty}^{\infty} c_m \exp[i(k_m - k_i)\beta\pi] \qquad (76)$$

in which according to (60) and (70)

$$k_m = (\varpi + 2m)a, \; k_i = \omega\nu/c \sin\theta_i \qquad (77)$$

with 0 ≤ ß < 1 since 0 ≤ x ≤ π/a.

So, taking into account (60), the granular periodic semi-infinite material behaves as a diffraction grating: the beam of plane waves propagating in the directions defined by the wave vectors with components (k_z, k_m) have their amplitudes modulated by the coefficients $c_m \exp(-ik_i \beta\pi)$. And, acccording to (76), the relations (74) and (75) become

$$A_i + A_r = A_t \Omega(\beta\pi),\ \nu/\mu \cos\theta_i (A_i - A_r) = \gamma A_t \Omega(\beta\pi),\quad 0 \leq \beta < 1 \qquad (78)$$

from which we get in terms of the incident amplitude Ai

$$A_r = -(\gamma\mu - \nu\cos\theta_i)(\gamma\mu + \nu\cos\theta_i)^{-1} A_i,$$
$$A_t = 2\nu\cos\theta_i (\gamma\mu + \nu\cos\theta_i)^{-1}\,\Omega^{-1}(\beta\pi)\, A_i \qquad (79)$$

So, the amplitude At is not constant on the interval (0, π/a).

4. DISCUSSION

The relation (6), leads to a consistent formalism but further work is needed to prove or to amend it. In any case, two different modes of harmonic plane waves propagate in these chiral materials. The Post constitutive relations used to characterize such media, allow to get exact analytic expressions for the amplitude of the electromagnetic field in each mode, a note-worthy property due, as noticed in the introduction, to the covariance of Post's relations under the proper Lorentz group. An excellent review of chiral nano-technology may be found in [17] with a discussion of two topics: nanoscale approaches to chiral technology and, corresponding to the situation considered here, nanotechnology that benefits from chirality. In particular, a section is devoted to chiral carbon nanotechnology and the authors conclude "possible applications of such materials in the field of biomedecine and biotechnology range from prepara-tion of novel antibacterial, cyclotonic and drug delivery agents to catalysis and materials science applications".

Remark: The analysis of Section 2 may be performed in left-handed chiral materials with negative ε, μ: just change ε, μ into $-|\varepsilon|, |-\mu|$

Granular periodic materials are currently used in mechanical engineering and, with the ob-jective to appraise their properties, theoretical studies have been devoted to acoustic wave propagation in these structures [18]. In electrical engineering, photonic crystals [19,20] are the main illustration of periodic nanomaterials and they take an increasing importance in today technology. But, they are not composite with inclusions immersed in a dielectric structure. For instance, a one-dimensional photonic crystal with a permittivity periodic in the direction of propagation may be described by an expansion in which U is the unit step function

$$\varepsilon(z) = \varepsilon_1 \sum_n [U(z - 2na) - U(z - \{2n+1\}a)] + \varepsilon_2 \sum_n [U(z - \{2n+1\}a) - U(z - \{2n+2\}a)] \quad (80)$$

and, the solutions of Maxwell's equations are the Bloch functions $\sum_m c_{k,m} \exp(ikz + 2i\pi mz/a)$ to be compared with (59) (and (80) with (49)). Incidently, (80) has a simple expression in terms of the square-sine function

$$\varepsilon(z) = \varepsilon + \rho \sin(az)/|\sin(az)|\ U(z),$$
$$\varepsilon_1 = \varepsilon + \rho,\ \varepsilon_2 = \varepsilon - \rho \quad (81)$$

which suggests to work with the Laplace transform of Maxwell's equations since tanh(πp/2a) is the Laplace transform of the square-sine function [21]. People fluent with the Laplace transform, could think in terms of p instad of z as they use to do with w instead of t.

In opposite to photonic crystals, composite granular materials with a continuous filling factor have no lattice structure and, as shown in Section 3.3, they rather behave as a smooth dielectric grating [22]. Some of the restrictive assumptions on the filling factor f could be somewhat released at the expense of more intricacy:

1) It would be interesting to check what happens when a higher order approximation than 0(f 2) is used;

2) When f(x) = f cos(2ax) is changed into f (x) = ∑0¥ fm(cos2max), the Mathieu equation becomes a Hill equation [14,15] with solutions similar to (55) but the recurrence relations bet-ween the coefficients cm is more intricate;

3) Finally a generalization to a two-dimensional filling factor f (x,y), periodic in x and y would approach more closely a real physical situation.

To sum up, the application of the Maxwell-Garnett theory to nano composites deserves further research, taking into account the innocuity or not of such materials in biomedecine [23]. This theory is also used to analyze, in the frame of surface plasmon polaritons, the scattering of TE, TM light waves from a composite material made of metallic nano spherical particles immersed inside a metallic structure such as Ag particles in a Sio2 matrix [24].

The 0(f 2) Maxwell-Garnett approximation of the periodic permittivity in the nanodoped medium of Section 3 implies that TE, and TM fields are solutions of the Mathieu equation as if they were diffracted from a dielectric grating [25].

REFERENCES

1. M. Khan, A. K. Sood, F. L. Deepak and C. N. R. Rao, "Nanorotors Using Asymmetric in Organic Nanorods in an Optical Trap," Nanotechnology, Vol. 17, No. 11, 2006, pp. S287-S290.

2. X. M. Guo, C. Jiang and T. S. Shi, "Prepared Chiral Nanorods of a Cobalt," Inorganic Chemistry, Vol. 46, No. 12, 2007, pp. 4766-4768.

3. J. C. Maxwell-Garnett, "Colours in Metal Glasses and Metallic Films," Philosophical Transactions of the Royal Society of London A, Vol. 203, No. 359-371, 1904, pp. 385-420.

4. C. A. Grimes, "Calculation of the Effective Electromagnetic properties of Granular Materials," in: A. Lakhtakia, Ed., Essays on the Formal Aspects of Electromagnetic Theory, World Scientific, Singapore, 1993, pp. 699-746.

5. L. Tsang and J. A. Kong, "Scattering of Electromagnetic Waves: Advanced Topics," Wiley Series in Remote Sensing, Wiley, New York, 2001.

6. P. Mallet, C. A. Guerin and A. Sentenac, "Maxwell-Garnett Mixing Rule in the Presence of Multiple Scattering," Physical Review B, Vol. 72, No. 1, 2005, p. 014205.

7. B. Shanker and A. Lakhtakia, "Extended Maxwell-Garnett Model for Chiral-in-Chiral Composites," Journal of Physics D: Applied Physics, Vol. 26, No. 10, 1993, pp. 1746- 1758.

8. Lakhtakia, V. K. Varadan and V. V. Varadan, "On the Maxwell-Garnett Model of Chiral Composite," Journal of Materials Research, Vol. 8, No. 4, 1993, pp. 917-922.

9. E. J. Post, "Formal Structure of Electromagnetics," NorthHolland Publications, Amsterdam, 1962.

10. Lakhtakia,V. K. Varadan and V. V. Varadan, "TimeHarmonic Electromagnetic Fields in Chiral Media," Springer, Berlin, 1989.

11. S. Bassiri, C. H. Papas and N. Engheta, "Electromagnetic Wave Propagation through a Dielectric-Chiral Interface and through a Chiral Slab," Journal of the Optical Society of America A, Vol. 5, No. 9, 1988, pp. 1450-1459.

12. M. Born and E. Wolf, "Principles of Optics," Pergamon Press, Oxford, 1965.

13. P. Hillion, "Light Beam Shifts in Total Reflection," Optics Communications,Vol. 266, No. 1, 2006, pp. 336-341.

14. E. T. Whittaker and G. N. Watson, "A Course of Modern Analysis," University Press, Cambridge, 1962.

15. Erdelyi, "Higher Transcendental Functions," McGrawHill, New York, Vol. 3, 1955.

16. P. Hillion, "Theoy of Electromagnetic Wave Propagation in a Longitudinally Periodic Cylinder," European Physical Journal B, Vol. 62, No. 4, 2008, pp. 477-480.

17. J. Zhang. M. T. Albelda, Y. Liu and J. W. Canary, "Chiral Nanotechnology," Chirality, Vol. 17, No. 7, 2005, pp. 404-420.

18. Inserra, V. Tournat and V. Gusev, "A Method of Controlling Wave Propagation in Initially Spatially Periodic Media," Europhysics Letters, Vol. 78, No. 4, 2007, p. 44001.

19. J. D. Joannopoulos, R. D. Meade and J. N. Wynn, "Photonic Crystals," University Press, Princeton, 1995.

20. Lopez, "Material Aspects of Photonic Crystals," Advanced Materials, Vol. 15, No. 20, 2003, pp. 1679-1704.

21. B. Van Der Pol and H. Bremmer, "Operational Calculus," Academic Press, Cambridge, 1959.

22. M. Neviere and E. Popov, "Light Propagation in Periodic Media," Marcel Dekker, Basel, 2005.

23. Poland, R. Duffin, I. Kinloch, A. Maynard, et al., "Carbon Nanotubes Introduced into the Abdominal Cavity of Mice Show Asbestos-Like Pathogenicity in a Pilot Study," Nature Nanotechnology, Vol. 3, 2008, pp. 423-428.

24. S. Kawata, "Near-Field Optics and Surface Plasmon Polariton," Springer, Berlin, 2001.

25. N. Yamaguchi, "High Order Bragg Diffraction by Dielectric Grating," Electronics and Communications in Japan, Vol. 69, No. 8, 2007, pp. 112-121.

7

CHAPTER

INFLUENCE OF THE CHEMICAL COMPOSITION OF COMPLETION FLUIDS ON THE PROPAGATION OF ELECTROMAGNETIC WAVES WITHIN OIL WELLS

Alexandre Ashade Lassance Cunha, Marco Aurélio Pacheco, José Ricardo Bergmann

Departamento de Engenharia Elétrica, Pontifícia Universidade Católica do Rio de Janeiro, Rio de Janeiro, Brazil

ABSTRACT

The propagation of electromagnetic waves in the annular region of oil wells was studied. The present study aims to analyse the propagation attenuation along the well, as well as the input impedance determined by a source placed near the wellhead. A coaxial waveguide model was adopted with heterogeneous dielectrics and losses. First, a wave equation solution for the waveguide is presented, assuming a homogeneous medium with losses, by solving the equation in cylindrical coordinates using the vector potential technique. An uncertainty analysis model is then developed to model the heterogeneous characteristics of the medium. Monte Carlo simulations were performed with the created model using data gathered from the literature. The results of the simulations indicate that propagation in the transverse electromagnetic mode has the smallest attenuation and that for depths of up to 4000 m, there is an attenuation of less than 52 dB. Furthermore, the input impedance ranges from 10 Ω to 10 kΩ because of the uncertainties involved in the problem in question.

Keywords: Electromagnetism; Oil Well; Wireless Telemetry; Monte-Carlo; Uncertainties; Heterogeneous Media

1. INTRODUCTION

Oil wells today are extremely complex and demand very expensive maintenance [1]. Modern drilling techniques can reach a few kilometres in depth, and the costs of a floating platform at open sea can cost billions of Brazilian Reais. Thus, it is critical to determine the internal conditions of the well, and indicators such as temperature, pressure and salinity must be constantly monitored [2].

Reliable telemetry by cables is very difficult to obtain due to the extreme conditions of the internal environment of an oil extraction well. As an example, there is continuous abrasion by sand and dirt that are carried by fluid flow. For this reason, the cables are periodically damaged and require replacements. Such replacements hinder the oil extraction process and, moreover, increase the costs due to the large amount of cables needed over the life of a single well.

The most obvious alternative to cabling is the adoption of wireless telemetry. However, the amount of electricity required to power wireless communication is not practical: power cables are needed because the use of batteries would lead to frequent maintenance, as they need to be recharged [1]. It is obvious, therefore, that the objective (and greater challenge) is to build a system without wires or batteries; that is, the downhole sensors must be powered and communicate with the base without the use of cables.

Several approaches are possible for wireless telemetry. One possibility would be the use of signal transmission techniques using a magnetic field [3]. In this method, a coil that is capable of inducing alternating current through the production pipe is used. This coil has sufficient intensity to transmit both power and the signal itself between the sensor and the base [4]. Through a second coil positioned at the sensor location, it is possible to recover the signal and power needed to feed and produce bilateral communication. However, the technique has a weakness: the production tube is not a transformer core, i.e., it is not designed to minimise the magnetic flux losses. Its hollow characteristic confers much loss by parasitic Foucault currents concomitantly with hysteresis losses, the cause of which is due to the inadequacy of the production tube material for magnetic purposes [5]. Therefore, for long distances, on the order of 3000 m or more, the method is unfeasible.

Another possible approach would be to analyse the oil well from the perspective of a coaxial propagation structure, which is formed by a conductor tube with a perfect centre (production tube) and a cylindrical shell that is coaxial with the tube and is also a perfect conductor. This approach makes it possible to conduct the analysis as it would be performed on a coaxial cable, and therefore, propagation occurs in the transverse electromagnetic mode [6]. This approach, however, has been studied without considering fluids of significant conductivity or possible uncertainties caused by temperature and electrical parameter variations of the fluid that fills the well.

The modelling difficulty concerns the fluid that fills the well. This fluid is heterogeneous because it exhibits significant variation in electric permittivity along the well depth. Furthermore, the actual fluid concentration varies from well to well, and at times within the well itself, which is another source of heterogeneity.

In this context, the present article proposes to study the electromagnetic propagation inside the annular area of oil wells. The goal is to develop a model that permits an approximation of the behaviour of the TEM propagation mode in the coaxial structure with losses and the previously cited heterogeneous characteristics. The first step is the deduction of the electric and magnetic field equations within the well, assuming a typical well structure. Thereafter, statistical models of the parameters that compose the medium are developed, considering their temperature variations and propagation frequency. Finally, the study uses Monte Carlo analysis to illustrate how the propagation attenuation behaves and to analyse the input impedance of the coaxial waveguide as a function of position.

2. MODELLING

2.1. Well Definition

The well will be modelled according to Figure 1. This figure depicts a coaxial guide bounded by perfect conductor metals with a homogeneous dielectric and losses. The upper extremity is completely closed by a perfect conductor metal to model the valve and metal duct assembly present in a wellhead. The bottom of the well is composed of concrete; however, this fact is neglected, and the well is assumed to behave as a semi-infinite waveguide.

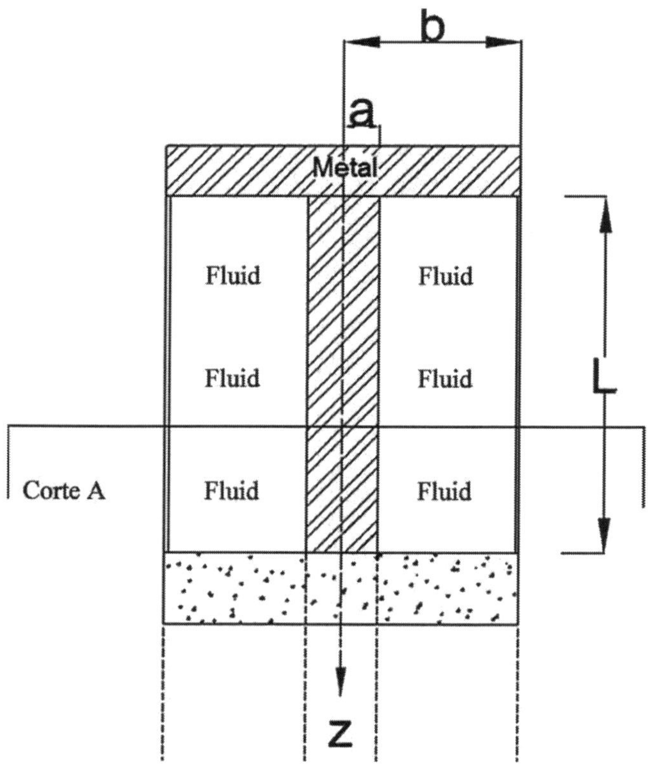

Figure 1. Adopted well structure. A semi-infinite well is assumed and is formed by two concentric cylinders and a metal block covering the top of the waveguide.

2.2. Deduction of the Propagation Equations

Modelling is performed using Maxwell's equations and the concept of vector potentials. The Maxwell's equations used are Equations (1) and (2):

$$\nabla \times \boldsymbol{E} = -\boldsymbol{M}_i - j\omega \boldsymbol{B} \tag{1}$$

$$\nabla \times \boldsymbol{H} = \boldsymbol{J}_i + \sigma_e \boldsymbol{E} + j\omega\varepsilon' \boldsymbol{E} \tag{2}$$

The real number constant ε' is the real part of the complex electrical permittivity of the medium. Furthermore, the complex constant σ_e is called effective conductivity and it represents the linear relation of conduction current and electric field on the medium.

Consider a medium with no source of magnetic charge $(\nabla \cdot \boldsymbol{B} = 0)$. Because $\nabla \cdot \nabla \times = 0$, the curl of A can be defined as

$$\nabla \times \boldsymbol{A} \equiv \boldsymbol{B} = \mu \boldsymbol{H} \tag{3}$$

Substituting 3 in 1 and considering $M_i = 0$ (Induced magnetic flux equal to zero), we obtain:

$$\nabla \times \boldsymbol{E} = -\nabla \times j\omega \boldsymbol{A} \tag{4}$$

Or:

$$\nabla \times \left(\boldsymbol{E} + j\omega \boldsymbol{A}\right) = 0 \tag{5}$$

Since $\nabla \times \nabla = 0$, an electric scalar potential g_e can be defined such that:

$$\boldsymbol{E} + j\omega \boldsymbol{A} \equiv -\nabla g_e \tag{6}$$

Applying the identity $\nabla \times \nabla \times = \nabla(\nabla \cdot) - \nabla^2$ in Equation (3) and using Equation (2), we obtain:

$$\begin{aligned} &-\nabla\nabla\cdot\boldsymbol{A}+\nabla^2\boldsymbol{A} \\ &=-\mu\boldsymbol{J}_i+\mu\left(\sigma_e+j\omega\varepsilon'\right)\left(\nabla g_e+j\omega\boldsymbol{A}\right) \\ &\sigma_e\equiv\omega\varepsilon''+\sigma_s \end{aligned} \tag{7}$$

In Equation (7), the parameter σ_e is the effective conductivity of the medium, while σ_s represents the static conductivity, that is, the conductivity when frequency is zero. At this moment, we are in a position to define $\nabla\cdot\boldsymbol{A}$. To simplify Equation (7), we use the gauge:

$$\nabla\cdot\boldsymbol{A}\equiv-\frac{\gamma^2}{j\omega}g_e \tag{8}$$

where $\gamma_1^2\equiv j\omega\mu\left(\sigma_e+j\omega\varepsilon'\right)$ is the complex constant of propagation of the medium. This step simplifies the equation for the electrical potential, resulting in:

$$\nabla^2\boldsymbol{A}=-\mu\boldsymbol{J}_i+\gamma^2\boldsymbol{A} \tag{9}$$

For homogeneous media, Equation (9), when solved, determines the electric vector potential in the medium, which can be used to obtain the equation for the electric field as a function of the electric vector potential:

$$\boldsymbol{E}=\frac{j\omega}{\gamma^2}\nabla\nabla\cdot\boldsymbol{A}-j\omega\boldsymbol{A} \tag{10}$$

We seek the solution in transverse electromagnetic mode (TEM), which is generated using $\boldsymbol{A}=\hat{a}_z A_z$ with the restriction $E_z = 0$, which leads to the following fields in cylindrical coordinates:

$$E_\rho=\frac{j\omega}{\gamma_1}\frac{A}{\rho}\sinh\left(\gamma_1 z\right) \tag{11}$$

$$H_\phi=-\frac{1}{\mu}\frac{A}{\rho}\cosh\left(\gamma_1 z\right) \tag{12}$$

$$\gamma_1^2 \equiv j\omega\mu\left(\sigma_e + j\omega\varepsilon'\right) \tag{13}$$

Note that because $E_\Phi = E_z = 0$, the boundary conditions that require a null tangential component in the metal extremities are already met. As the well has a metal structure at one of the extremities, one must also ensure that E_ρ $(z = 0) = 0$. Note, however, that this condition is also already guaranteed.

The above solution represents a propagation model in TEM mode in a coaxial medium with losses inherent to the medium inside. However, it is important to remember that the solution was obtained by assuming a homogeneous propagation medium.

2.3. Statistical Modelling of the Medium Constituent Parameters

The propagation medium in the well consists of an oilbased dielectric fluid composed of water, oil and salts (typically $CaCl_2$) [7]. This fluid is the centre of all electromagnetic propagation, and therefore, its electrical characteristics must be examined in detail. A study over the range of 1 MHz to 100 MHz was previously conducted [7], revealing significantly variable behaviour based on the chemical composition of the fluid.

To model the variation of conductivity with frequency, the effective conductivity concept is applied [8] using the following formula:

$$\sigma_e \equiv \sigma_s + \omega\varepsilon'' \tag{14}$$

From the experimental curves obtained in [8], the parameters σ_s and ε'' can be calculated using a least squares method.

An analogous model can be created to model the variation in frequency of the real relative permittivity with the frequency. With

$$\varepsilon' \equiv \varepsilon_s + \kappa\omega \tag{15}$$

and using the experimental curves in [8], it is possible to estimate the parameters and κ using a least squares approach.

In addition to the variation in frequency, variation in the medium ε_s constituent parameter can also be observed in temperature. Such variation cannot be neglected because inside a 5000 m deep well, it is impossible to ensure that the temperature is uniform along its entire length. This fact, therefore, characterises non-homogeneity along the z direction.

To circumvent the situation, a model that assumes an average temperature in the medium is adopted. This temperature, in turn, is considered constant throughout the well, which implies a homogeneous medium. Thus, the influence of the variation of this average temperature on signal propagation in the well can be analysed, assuming a valid range of average temperatures.

Mathematically, a coefficient of correction in temperature is defined as the ratio between the temperature value in question and the value of the reference temperature, here defined as 25°C. Thus,

$$\theta \equiv \frac{\sigma_e(\theta)}{\sigma_e(25^\circ\text{C})} \tag{16}$$

$$\theta \approx a_1\theta^2 + a_2\theta + a_3 \tag{17}$$

$$\sigma_e = \sigma_e(25^\circ\text{C}) \times \theta \tag{18}$$

The quadratic form was selected to present the best interpolating results using the experimental data from [8].

Similarly, there is increasing variation in the relative permittivity according to the temperature, as demonstrated by [7]. Again, it is used a quadratic model similar to the effective conductivity variation in temperature:

$$\theta_\varepsilon \equiv \frac{\varepsilon'(\theta)}{\varepsilon'(25^\circ\text{C})} \tag{19}$$

$$\theta_\varepsilon \approx a_1\theta^2 + a_2\theta + a_3 \tag{20}$$

$$\varepsilon' = \varepsilon'(25^{\circ}\mathrm{C}) \times \theta_{\varepsilon} \qquad (21)$$

3. EXPERIMENTS AND RESULTS

In all the experiments, a well with a length of 5000 m, an inner radius of 0.05 m and an outer radius of 0.1 m is assumed. All the Monte Carlo analyses were conducted with at least 1 million samples. The objective of the experiments was to obtain graphs and numerical values for the input impedance of the "oil well" waveguide and to analyse the propagation loss in the medium for 5000 m of depth.

All analyses were conducted over the range of 1 MHz to 100 MHz. For the other free parameters, the modelling used random variables whose distribution was selected to reflect their most common values. Table 1 summarises the values selected for each free parameter of the previously described model and their respective distributions. For simplicity, when the random variable has simple and obvious domain restrictions, a uniform distribution was selected; otherwise, a normal distribution was used. Furthermore, statistical independence was assumed between the variables.

First, an experiment was conducted to calculate the attenuation in the well. The attenuation is directly dependent on the real part of the constant of propagation, whose square is defined as $\gamma_1^2 \equiv j\omega\mu(\sigma_e + j\omega\varepsilon')$.

The models proposed in Section 2.3 were used for the conductivity and electric permittivity, and the relative magnetic permeability was assumed to be equal to 1.

The Figure 2 presents a boxplot of the constant of attenuation for the various frequencies between 1 and 100 MHz. Values that appear in the graph as outliers are from highly unlikely combinations of the random variables of the problem and are most likely not physically feasible and, therefore, must be disregarded. The height of the boxes represents the range of values most likely to be observed, and its average point is a good approximation of the average. The boxplot of Figure 3 demonstrates that the attenuation constant increases with an increase in frequency, as expected. Furthermore, note that the deviation of the coefficient increases with the elevation of the propagation frequency, which makes the system design even more difficult. Thus, it is clear that lower frequentcies are desirable from the point of view of signal attenuation.

Table 1. Random variables used for modelling uncertainties. For simplicity, when the random variable has simple and obvious domain restrictions, a uniform distribution was selected; otherwise, a normal distribution was used.

Variable	Average/a	Deviation/b	Distribution
σ_z (S/m)	0	4.0×10^{-5}	Uniform
ε_z (F/m)	3	12	Uniform
ε^{*} (F/m)	1.0×10^{-13}	3.0×10^{-12}	Uniform
k (F/H$_z$·m)	2.5×10^{-9}	5.0×10^{-10}	Normal
θ_{med} (°C)	40	5	Normal
a_1 (*permissivity*)	9.0×10^{-4}	2.0×10^{-4}	Normal
a_2 (*permissivity*)	-3.0×10^{-12}	6.0×10^{-3}	Normal
a_3 (*permissivity*)	1.2	0.1	Normal
a_1 (*permissivity*)	1.0×10^{-5}	2.0×10^{-6}	Normal
a_2 (*permissivity*)	0.0	5.0×10^{-4}	Normal
a_3(*permissivity*)	1.0	1.0×10^{-2}	Normal

Using the TEM mode equations presented in Equation (2), it is clear that the wave propagating in the direction of the bottom of the well has power attenuation given by $A = \exp\left(\alpha^2 L^2\right)$. Using the statistical analysis of the coefficient of attenuation, Table 2 was generated.

The values from the table represent the upper limit of attenuation for 95% of the cases. Thus, for example, for a depth of 1000 m, P (A ≤ 13 dB) = 0.95. The results presented in Table 2 do not agree with the propagation for L = 5000 m of depth because of the high-energy attenuation (130.3 dB). However, it is essential to note that the table represents an upper limit of

Table 2. Estimated attenuation of electromagnectic waves propagating in the annular region of the well.

L (m)	Electric field attenuation (dB)	Power attenuation (dB)
1000	13.0	26.1
2000	26.1	52.1
3000	39.1	78.2
4000	52.1	104.2
5000	65.1	130.3

attenuation, taking 95% of the possible combinations of propagation medium and an average temperature. Moreover, as the table demonstrates, propagations up to 2000 m depth are highly acceptable, as an attenuation of up to 52.1 dB is observed in current communication technology. Even at greater depths, the possibility of communication cannot be excluded.

The second experiment was performed to analyse the input impedance of the coaxial waveguide as a function of the excitation source position. The experiment was conducted at 1 MHz.

Setting the impedance at any point as the ratio E_ρ / H_ϕ, we obtain

$$Z = Z_0 \tanh(\gamma_1 z) \tag{22}$$

$$-L < z \le 0 \tag{23}$$

$$Z_0 = -\mu \frac{j\omega}{\gamma_1} \tag{24}$$

The position selected for analysis was at $1/4$ wavelength from the wellhead, that is, $Z = \lambda/4$. However, the wavelength itself is a random variable because $\lambda = 2 * \pi/\beta$ and $\beta = \mathrm{Im}\{\gamma\}$ is a random variable. Monte Carlo analysis revealed that λ varies between 90 m and 170 m for 90% of the cases, leading to the selection of position $Z = ((90+170)*1/2)/4 = 32.5$ m.

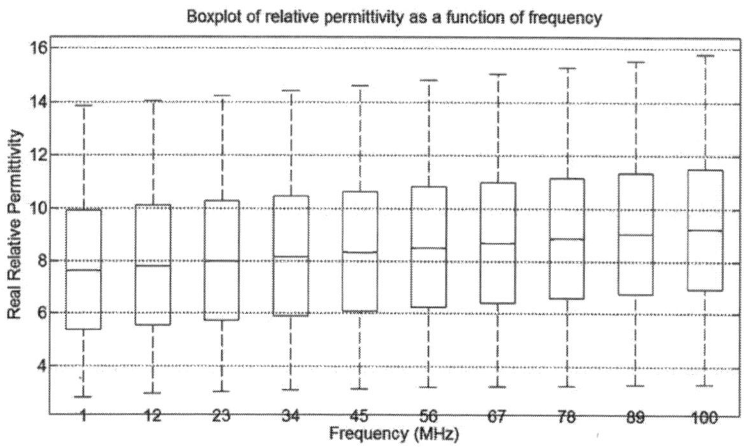

Figure 2. Boxplot of relative permittivity as a function of frequency. The horizontal axis represent the frequency of propagation in MHz, while the vertical axis represent the actual value of the electrical permittivity.

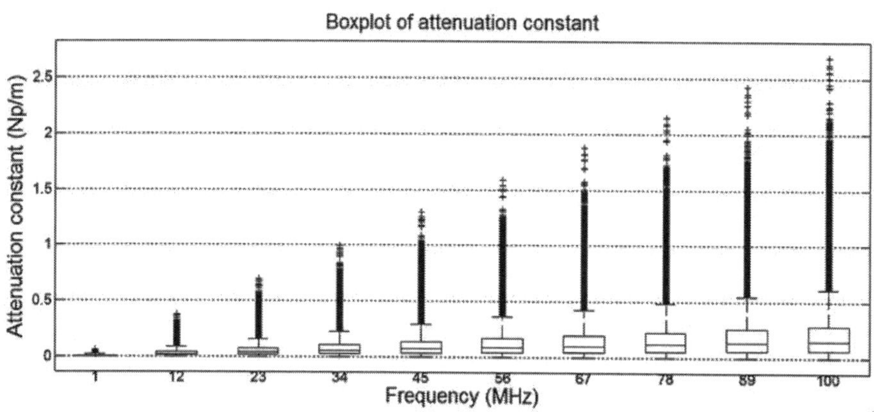

Figure 3. Boxplot of attenuation constant as a function of frequency.

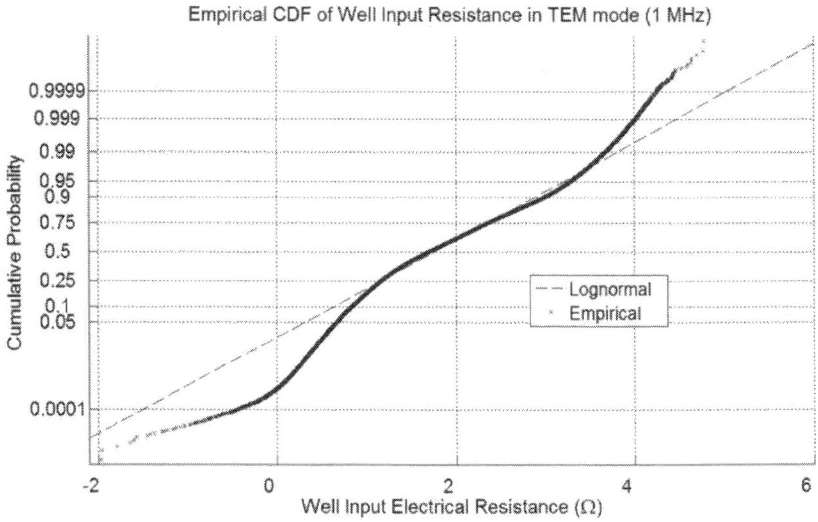

Figure 4. Empirical cumulative distribution function of the well input resistance in TEM mode at 1/4 wavelength from well head. The horizontal axis show the log on base 10 of the value of input resistance in Ohms, while the vertical axis show the cumulative distribution function value.

Using Monte Carlo analysis again, it can be observed in Figure 4 that the input resistance (real part of the impedance) of the system exhibits great variation, ranging from 10 Ω to 1.0 kΩ for 90% of the cases. This variation is due to the variation of the relative permittivity in the medium. Therefore, it is necessary to design a generator circuit that provides a good match for a wide range of input impedances, i.e., the generator/receiver circuit must lose as little power as possible by reflection.

4. CONCLUSIONS

The present study focused on analysis of the electromagnetic propagation inside the annular area of oil wells, assuming the interior medium is composed of dielectrics with significant conductivity. The well behaviour was evaluated with respect to the input impedance and propagation attenuation.

To quantify the well behaviour with respect to electromagnetic propagation, a coaxial waveguide model was developed, modelling the wellhead as a metal cap that completely closes one extremity of the waveguide. First, we solved the wave equation for a homogeneous cylindrical coaxial waveguide, resulting in an analytical model. Then, an uncertainty model was adopted to approximate the heterogeneous and imprecise characteristics of the fluid that functions as a dielectric inside the well.

Using this model, two experiments were developed, both by simulation. The first aimed at analysing the input impedance observed by a source positioned at 1/4 of the wavelength from the wellhead, while the other aimed at analysing the attenuation as a function of length or well depth.

Due to the uncertainties present, the input impedance was observed to vary from 10 Ω to 1 kΩ for the frequency of 1 MHz. The conclusion was obtained by Monte Carlo simulation and was applied to the expression derived for the waveguide input impedance.

By performing the attenuation analysis, it was concluded that for 95% of the cases, the constant of attenuation in TEM mode is less than 0.8×10^{-4} Np/m for a frequency of 1 MHz. The power attenuation at 4000 m depth was also observed to be approximately 100 dB for the same frequency of 1 MHz, with 95% probability. Although 100 dB appears to be a large attenuation, most wells in operation are between 1000 m and 2000 m in length, and the attenuation is much lower at these depths.

REFERENCES

1. S. Brilles, "Remote Downhole Well Telemetry," US Patent US6766141, 2004.

2. J. A. D. Rosa, A. J. Carvalho and R. D. S. Xavier, "Petroleum Reservoir Engineering," Rio de Janeiro, 2006.

3. F. Sakata, H. Wakiwaka, M. Hanabusa, N. Yamazaki and H. Yamada, "Performance Analysis of Long Distance Transmitting of a Magnetic Signal in a Cylindrical Steel Rod," IEEE Translation Journal on Magnectics in Japan, Vol. 8, No. 2, 1993, pp. 102-106.

4. F. Harold J. Vinegar, R. R. Burnett, G. C. W. M. Savage and J. W. Hall, "Permanent Downhole, Wireless, TwoWay Telemetry Backbone Using Redundant Repeaters," US Patent US6633236B2, 2003.

5. B. W. Kennedy, "Energy Efficient Transformers," New York, 1998.

6. K. A. Safynia and R. W. McBride, "System and Method for Communicating Signals in a Cased Borehole with Tubing," US Patent US4839644, 1989.

7. P. A. Patil, et al., "Experimental Study of Electrical Properties of Oil-Based Mud in the Frequency Range from 1 to 100 MHz," SPE Drilling and Completion, Vol. 25, No. 3, 2010, pp. 380-390. doi:10.2118/118802-PA

8. C. A. Balanis, "Advanced Engineering Electromagnetics, Vol. 52, No. 1," Wiley, Hoboken, 1989, p. 1008.

8
CHAPTER
FIELD ESTIMATION THROUGH RAY-TRACING FOR MICROWAVE LINKS

Vittoria Bosisio[1]

[1] National Research Council of Italy, CNR/IEIIT c/o Politecnico di Milano, Italy

1. INTRODUCTION

Engineers using microwave radio links have to address the effects of multipath propagation that arise when several rays arrive at the receiver after travelling along different paths from the transmitter. Rays, along their propagation, undergo reflections at the earth's surface or at variations in the refractive index or its gradient.

Since the '60s, standard ray theories for radio wave applications use effective earth radius concepts, ray bending based on Snell's law in a layered spherical atmosphere, or analytical method, limited to simple refractive index profiles (Du Castel, 1966; Livingston, 1970). At that time, scientists were interested in statistics of fading, distinguishing between *fast* or *slow fading* and the typical results were either nomograms of attenuation as a function of distance for a set of common frequency and height values or cumulative distribution functions representing the fraction of time that the signal was expected to be received at or below a given level (CCIR, 1978). Several techniques were employed to calculate the field strength, such as Cornu's spiral derived by the Huygens' principle together with the introduction of the specular and the diffuse reflection contribution (Hall, 1979). In that approach, the field amplitude was determined assuming the principle of conservation of energy, i.e. the flux of the energy along a ray is the same at every cross section of a narrow tube of rays (Keller, 1957).

Nowadays, radio communications live a renaissance due to the development of mobile and wireless communications and the increasing demand of wideband services. The need of greater rate of transmission is accomplished using higher frequencies for which effects like the tropospheric scatter can affect dramatically the quality of the received signal. Thus, aspects such as analysis of delay power spectra or of potential intersystem interferences require a more realistic modelling. As a matter of fact, local fluctuations in the refractive index n can cause scatter, while its abrupt changes with the height can cause reflection and originate ducting layers, in which the rays are reflected and refracted back in such a way that the field is trapped inside these layers (Bean & Dutton, 1968).

Although the goal of radio engineering is to transmit reliably from the source to the receiver, this is of no use if the received signal is unintelligible due to interference or if the transmission causes interference with other systems. So an analysis of the channel behaviour is mandatory to meet both link design and system requirements. Besides, the availability of geographic databases, digital elevation models and the increasing computing power let us face more realistic description to model the interaction of the propagating wave with the atmosphere and the surrounding scenario (Driessen, 2000; Kurner & Cichon, 1993; Lebherz et al., 1992).

In more recent developments, ray modelling of wave propagation addresses the dispersive effects of perturbed atmosphere on the performance of high-capacity digital radio channel (Akbarpour & Webster, 2005; Sevgi & Felsen, 1998).

Besides telecommunications applications, a proper modelling of atmospheric propagation is of concern in atmospheric sciences with particular focus on radio occultation data analysis for assimilation with numerical weather prediction models (Pany et al., 2001). Also, the decrease of propagation speed related to the atmospheric refractivity causes tropospheric delay, which influences applications of the Global Navigation Satellite System (Eresmaa et al., 2008).

The author presents a widely diffuse technique in the domain of seismic studies that accommodates lateral variations in the medium properties (Farra, 1993). As an asymptotic technique, based on high frequency approximation, it permits fast computation but provides a local solution of propagation problem (Červený et al., 1977). The technique was adapted for application with electromagnetic waves and specifically tailored for signals travelling in the atmosphere. The hypothesis of horizontal uniformity can be removed and no stratification is needed to calculate the ray trajectory. The terrain profile coordinates are mapped over a Cartesian reference through analytical transformation. Refractive index values are given in the same Cartesian reference. Hence, propagation is modelled in a two dimensional range-height scenario over irregular terrain through non-homogeneous atmosphere.

The field amplitude is evaluated by means of a perturbation technique using paraxial rays to observe the wave front structure along the path from the transmitter towards the receiver.

This approach allows to analyze system performance and the channel impulse response in presence of any kind of atmosphere, characterized by local values of refractive index n in the bi-dimensional panel including the antennae and the terrain profile. The variations of the refractive index n along the third dimension are taken equal to zero; nonetheless, the field amplitude is calculated correctly because of the paraxial approximation which takes place in a 3D domain. Median power strength of the numerical results was compared with predictions given by Friis' Formula (Balanis, 1996).

2. MODELLING

The proposed ray tracing technique is widely used in seismic for subsoil investigation and it follows an approach based upon the integration of the Eikonal equation with Hamiltonian-Jacobi technique (Kružkov, 1975). According to this formulation, rays are defined by the vector $y(\tau)=(x(\tau), p(\tau))$, where $x(\tau)$ and $p(\tau)$ are the position vector and the slowness (inverse of phase velocity) vector along the ray, both function of the integration variable τ and of the initial conditions (launching point and direction). The slowness vector p is defined as $\mathbf{k}/\omega$ and $e^{j\mathbf{k}\cdot\mathbf{x}}$ is the phase function. The vector $y(\tau)$ satisfies the Hamilton differential equations, whose solutions describe the wave propagation in the medium, under the asymptotic assumption (Abramovitz & Stegun, 1970):

$$\begin{cases} \dfrac{d\mathbf{x}}{d\tau} = \nabla_p \mathcal{H} \\ \dfrac{d\mathbf{p}}{d\tau} = -\nabla_x \mathcal{H} \end{cases} \tag{1}$$

where $_P$ and x represent the gradient computed versus the slowness vector and the position vector, respectively. H is the Hamiltonian function describing the wave propagation in the considered medium, i.e. the atmosphere, chosen as follows:

$$\mathcal{H}(x,p) = \frac{1}{2}\left(v^2(x)p^2 - 1\right) \tag{2}$$

being v (x) the propagation speed of the medium at the position vector x.

The advantage of such an approach is twofold: it takes into account the medium inhomogeneities that originate multipath propagation and wavefronts folding (caustics), and it permits to consider both vertical and long range variations of the atmospheric model, without any approximation like flat-earth model (Hall, 1979). Also, it permits to keep separated the different field contributions due to the different interactions between the rays and the surrounding scenario, which impacts on the phase calculation of the overall received field.

2.1. Ray tracing

A suitable discussion for ray tracing can be found in (Farra, 1993); nevertheless, in this section some basic concepts and analytical details are reported.

Substituting (2) in (1), one obtains the equations for the vector y(τ):

$$\begin{cases} \dfrac{dx}{d\tau} = pv^2(x) \\ \dfrac{dp}{d\tau} = -\dfrac{\nabla_x v(x)}{v(x)} \end{cases} \tag{3}$$

x and p define the 6D phase-space, where the solution of (3), i.e. the rays, are the characteristic lines of the Eikonal equation, thus enabling the interpolation without ambiguity on a single fold (Vinje et al., 1993). For this reason rays are also called bi-characteristics lines of the wave equation. By choosing as integration variable the ray propagation delay or travel time T, system (3) becomes:

$$\begin{cases} \dfrac{dx}{dT} = c^2(x)p_x \\ \dfrac{dy}{dT} = c^2(x)p_y \\ \dfrac{dz}{dT} = c^2(x)p_z \\ \dfrac{dp_x}{dT} = \dfrac{1}{c^2(x)}\dfrac{\partial c(x)}{\partial x} \\ \dfrac{dp_y}{dT} = \dfrac{1}{c^2(x)}\dfrac{\partial c(x)}{\partial y} \\ \dfrac{dp_z}{dT} = \dfrac{1}{c^2(x)}\dfrac{\partial c(x)}{\partial z} \end{cases} \tag{4}$$

The solution of system (4) describes, under the asymptotic assumption, the propagation in the atmosphere in terms of ray trajectories and time delay.

2.2. Amplitude calculation

Anomalous conditions such as medium inhomogeneity or velocity model variations can affect dramatically the wavefront propagation. Under these circumstances the computation based upon the spherical divergence assumption results in erroneous evaluations.

One possible choice is to compute the amplitude through paraxial rays (first order perturbation theory), used to determine the flow tube, whose deformations, together with the theorem of the conservation of the flux along the ray path, allow the calculation of the amplitude A at the time τ.

Let us consider a reference ray with characteristic vector $y_0(\tau)=(x_0(\tau), p_0(\tau))$. A paraxial ray is obtained from the reference one by applying the first order perturbation theory, so paraxial rays coordinates are defined by:

$$x(\tau)=x_0(\tau)+\delta x(\tau)$$
$$p(\tau)=p_0(\tau)+\delta p(\tau)$$

(5)

Tracing paraxial rays consists in finding in the phase-space the canonical perturbation vector $\delta y(\tau)=(\delta x(\tau),\delta p(\tau))$. These perturbations in the trajectory are due to small changes in the initial conditions of the ray.

The linear system for the calculation of paraxial rays is obtained by inserting the perturbation vector given by (5) in system (1) and developing to the first order:

$$\frac{d\delta y}{d\tau}=\mathbf{A}(\tau)\cdot\delta y$$

Where

$$\mathbf{A}=\begin{bmatrix} \nabla_{\mathbf{x}}\nabla_{\mathbf{p}}H & \nabla_{\mathbf{p}}\nabla_{\mathbf{p}}H \\ -\nabla_{\mathbf{x}}\nabla_{\mathbf{x}}H & -\nabla_{\mathbf{p}}\nabla_{\mathbf{x}}H \end{bmatrix}$$

(7)

is a 6 6 matrix whose elements are the derivatives of the Hamiltonian function calculated on the reference ray.

Paraxial rays allow to define the flow tube, schematically represented in figure 1.

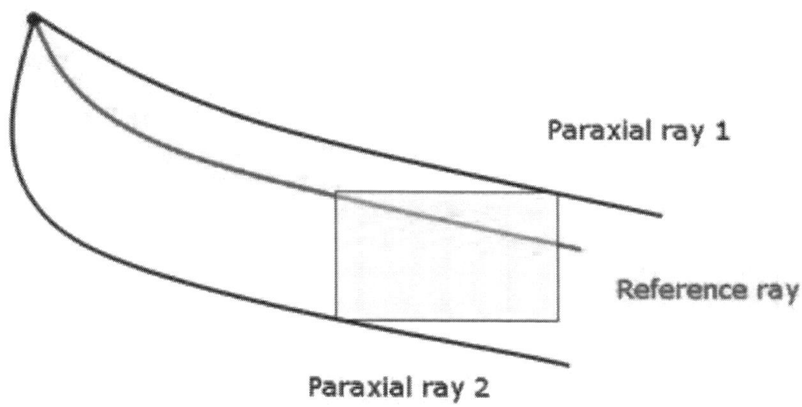

Figure 1. Flow tube definition through paraxial ray tracing

The deformation of the flow tube along the ray path ℓ is given by the jacobian J:

$$J = \frac{\partial(x,y,z)}{\partial(\tau,\vartheta,\phi)} = \det\begin{bmatrix} \frac{\partial x}{\partial \tau} & \frac{\partial x}{\partial \vartheta} & \frac{\partial x}{\partial \phi} \\ \frac{\partial y}{\partial \tau} & \frac{\partial y}{\partial \vartheta} & \frac{\partial y}{\partial \phi} \\ \frac{\partial z}{\partial \tau} & \frac{\partial z}{\partial \vartheta} & \frac{\partial z}{\partial \phi} \end{bmatrix} \tag{8}$$

where τ is the sampling parameter, θ the elevation angle and ϕ the azimuth angle. The amplitude of the plane wave associated to the ray is computed considering both the deformation of the flow tube and the conservation of power density flow law along the ray path. The volume element of the flow tube can be expressed as $dV = J\, d\tau\, d\vartheta\, d\phi$ or $dV = dS\, d\ell$.
Thus, the elementary surface dS has the following expression:

$$dS = \frac{dV}{d\ell} = \frac{J\, d\tau\, d\vartheta\, d\phi}{d\ell} = \frac{J}{v} d\vartheta\, d\phi \tag{9}$$

From the conservation of the power density flow along the ray (Keller, 1957) and known the amplitude A_i at τ_i, it follows that the amplitude A_{i+1} at τ_{i+1} is:

$$A_{i+1} = A_i \frac{v_{i+1}}{v_i} \sqrt{\frac{J_i}{J_{i+1}}} \tag{10}$$

Equation (10) shows that the ray amplitude depends on the velocity model. Hence, paraxial rays take into account anomalous propagation conditions.

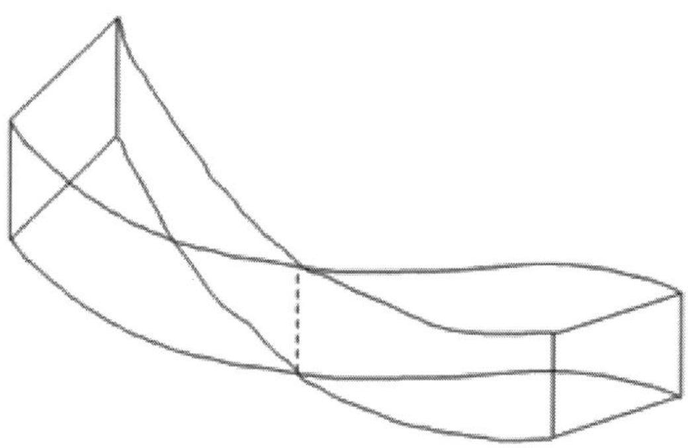

Figure 2. Sketch of a caustics of the first order: the flow tube looses one dimension

The singularity of the Jacobian J is a singularity, known as caustic, of the plane wave associated to the ray (Kravtsov & Orlov, 1990). Caustics arise when the ray field folds. Caustics are of first order when the flow tube looses one dimension and of the second order when two are the dimensions lost. The wave front folding affects the phase of the field carried by the wave with a phase shift of $\pi / 2$ radians for each caustic order. These events are carefully accounted so that the proper phase shifts can be applied to the field.

2.3. Parameterization

The modelling is performed in a 2D panel, as far as the terrain profile and the atmosphere characteristics are concerned, while the ray trajectories are computed in a 3D space.

This assumption is based on the hypothesis that lateral variations in the propagation medium and in the terrain profile are negligible in the third dimension, which does not strictly holds.

Therefore, the propagation occurs in the so called Earth system (s,h), where s is the range measured along the idealized spherical earth and h is the altitude taken along radial direction passing through the Earth centre, respectively. Instead, for sake of simplicity, all calculation are developed in a Cartesian reference system (x,z) related to the Earth by the following transformation relationships:

$$x = (R_0 + h)\sin\left(\frac{s}{R_0}\right)$$

$$z = (R_0 + h)\cos\left(\frac{s}{R_0}\right) - R_0 \tag{11}$$

where R_0 is the Earth radius. The transformation from cartographic coordinates (s,h) to absolute coordinates (x,z) is given by the inverse transformation:

$$s = R_0 \arctan\left(\frac{x}{z + R_0}\right)$$

$$h = \sqrt{x^2 + (z + R_0)^2} - R_0 \tag{12}$$

being $\theta = s / R_0$ the angle at the centre of the Earth, as shown in figure 3.

This way to proceed allows to take into account the actual geometry of the problem without introducing approximations like equivalent Earth radius. Also, it keeps separated the domain in which the radio link characteristics are defined and the computational one, where rays are traced.

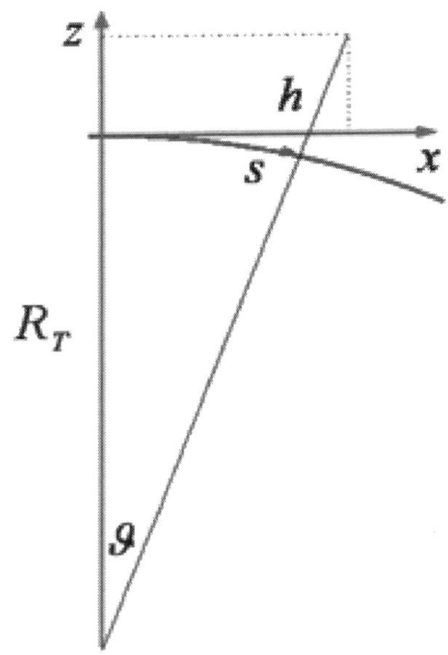

Figure 3. The two reference systems: cartographic coordinates (s,h) and absolute ones (x,z).

The characteristics of the atmosphere are taken into account in terms of its propagation velocity using several quantities that are generally function of the position vector x:

$$v(x) = \frac{c}{n(x)} = \frac{1}{p(x)} \tag{13}$$

where c is the speed of the light and n(x) is the refractive index of the atmosphere, which usually is expressed in terms of the refractivity N:

$$N = (n-1) \cdot 10^6 \tag{14}$$

N is a dimensionless quantity but in literature it is often measured in N-units or in parts per million (ppm). Its value depends upon the altitude and the range locally according to the atmospheric pressure P, the vapour pressure P_V and the temperature T as in the following (Bean and Dutton, 1968) :

$$N = 77.6\frac{P}{T} + 3.73 \cdot 10^5 \frac{P_V}{T^2} \tag{15}$$

Besides the variations in the refractive index, or its gradient, rays along their propagation undergo reflections and diffraction at the earth's surface. At the present stage of the modelling, neither the diffraction mechanisms nor any scattering features are included except reflection.

The trajectory of the reflected ray is given by the Snell's law under the hypothesis that locally the ground acts as a perfect smooth surface. The initial conditions with which the reflected ray is traced depend both on the geometric characteristics of the terrain profile and on the ray direction of incidence. The amplitude and the phase of the rays reflected from the ground are computed according to the chosen ground parameterization, i.e. the electric soil properties. These determine the value of the ground impedance η, which is a function of both soil permittivity ε_r and conductivity σ (Ulaby, 1999):

$$\eta = \sqrt{\frac{\mu}{\varepsilon_0 \varepsilon_r}\left(1 - j\frac{\sigma}{\omega \varepsilon_0 \varepsilon_r}\right)} \tag{16}$$

being ε_0 and ω the free space permittivity and the angular frequency respectively.

Once that the rays trajectories are computed, together with their complex amplitudes - weighted by the antenna pattern - and the travel times, they are classified into different wave fronts and then interpolated on the locations where the field is desired. The total field is obtained by addition of the single contribution of each wave front.

3. RECEIVED FIELD ESTIMATION

The field reconstruction is described by means of a pilot example of a 4.5 GHz point-to-point radio link 80 km long affected by multipath. The aim of the technique is to evaluate the vertical electric field (or the received power) intensity at the receiver range. According to the considered geometry, this leads to the prediction of the vertical profile of the electric field E(h) (or P(h)), where h is the receiver height ranging between

$$\left(h_0 - \Delta h / 2\right) \ \left(h_0 + \Delta h / 2\right)$$

. In the following we refer to this interval of width Δh as target zone or simply target.

The modelled atmospheric condition is indicated in the panel of figure 4, while figure 5 shows the vertical profile of N at few chosen distances. The N value is comprised between 280 and 360 N-units and it shows variations in both dimensions. This atmospheric model -taken from (Bean and Dutton, 1968), p.325 - let us focus on several kind of interactions and their effects on the propagating signal that could possible occur in actual conditions.

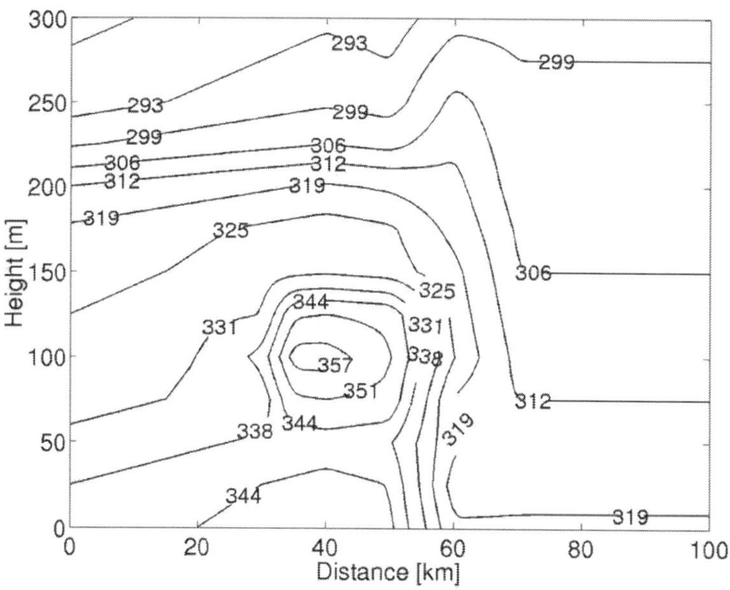

Figure 4. panel of the atmosphere charachteristics. Contour lines report the value of the refractivity index N

To appreciate the role of the variations in the refractivity, and without loss of generality, the ground profile was chosen as a plane surface characterized by the values of permittivity $\varepsilon_r = 3$ and conductivity $\sigma = 0.001$ [S/m] in the frequency band of interest.

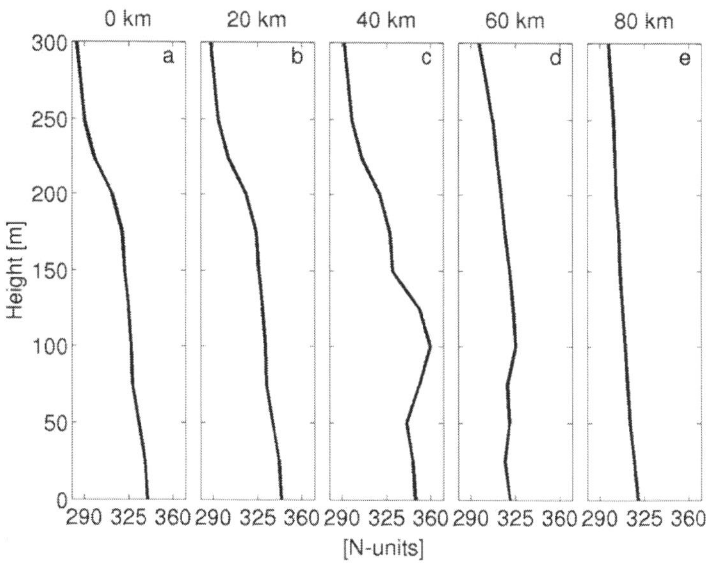

Figure 5. Vertical profile at 5 chosen range distances.

The top image in figure 6 shows the trajectories of the rays belonging to a chosen elevation angular sector (-0.5 ÷ 0.5 with respect to the local horizon) traced in the cartographic system. The transmitter is in the origin of abscissas axis at 100 m above the ground. Besides the wavefront that carries the direct arrival of the signal, one may observe that several wavefronts are involved, resulting from the inhomogeneities of the atmosphere and ground reflection.

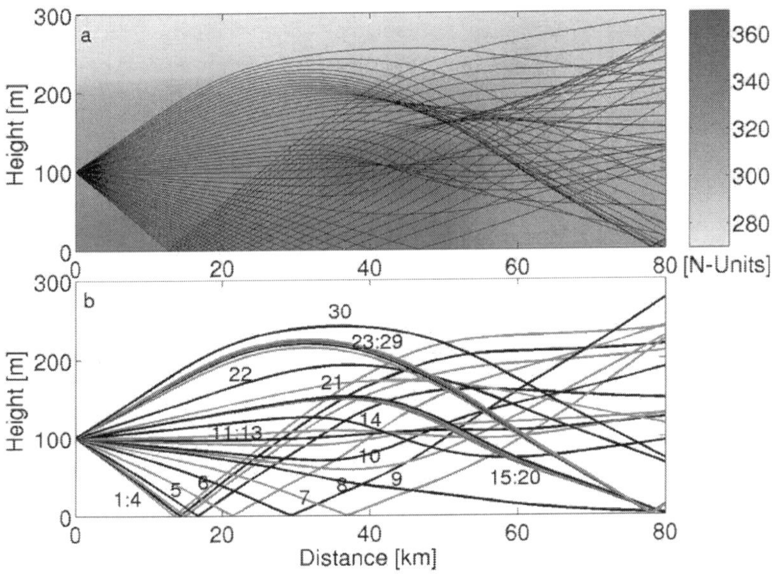

Figure 6. Ray pattern (top) and wavefronts (bottom)

The bottom image of figure 6 shows the rays pertaining to the 30 wavefronts that results from the ray tracing process. The first step consists in grouping rays in different wave fronts, each characterized by the same history such as same number and location of caustics or reflections. Wavefronts are numbered and rays belonging to the same wavefront can be used to interpolate the field. As a matter of fact, rays arrive at the receiver range at discrete values so that the vertical profiling of the field intensity involves three steps: wavefronts separation; interpolation of the ray comb belonging to the same wavefront; coherent addition of the wavefronts themselves.

3.1. Wave front separation

The first criterion of wavefront separation that could come in mind is that of the travel time arrivals. Figure 7a shows the multi-valued behaviour of the wavefront delay of arrivals at different heights, while figure 7b shows how the parameterization of the wavefronts in the ray launching angle resolves the ambiguity: working in the angle domain seems the natural way for unfolding multi-valued ray fields (Operto et al., 2000).

The wavefronts separation is based on a broad and on a narrow selection process applied consecutively. The broad selection process involves the ray history: as for example, two rays belong to two different wavefronts if they are reflected a different number of times. The narrow selection process uses as criterion the angular distance: if two rays are reflected the same number of times they belong to the different wavefronts when their elevation starting angles difference is greater than $\delta\theta$. This angular distance has the role of guarantee the proper accuracy in the interpolation of the ray field and it depends on the velocity model of the medium (Sun, 1992).

The results of the wavefront separation process are organized in a database whose structure is schematically represented in figure 8. Wavefronts and rays are organized in a record frame that gives for the kth wavefront the propagation delay τ_k, the amplitude A_k and the phase ϕ_k obtained by interpolation of the travel time T_{ik}, a_{ik} and φ_{ik} of the rays at their arrival height, where i pertains to the individual rays in the wavefront k.

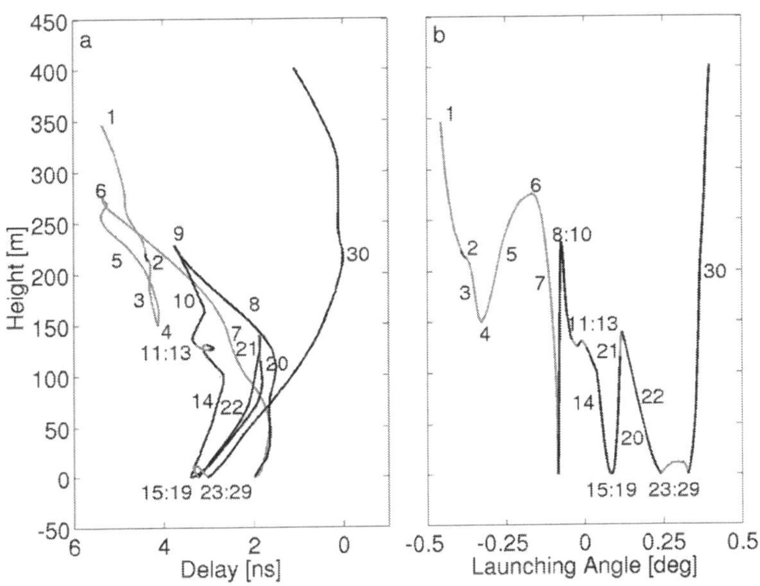

Figure 7. Wavefronts time delay dispersion (a) and launching angle distribution (b)

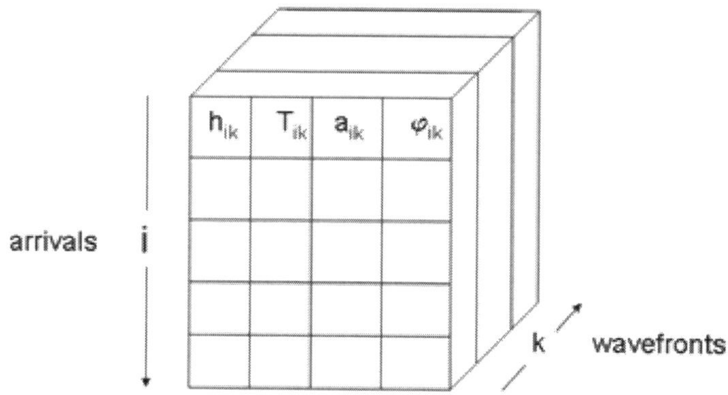

Figure 8. Ray field database: the arrivals are organized according to their wavefronts

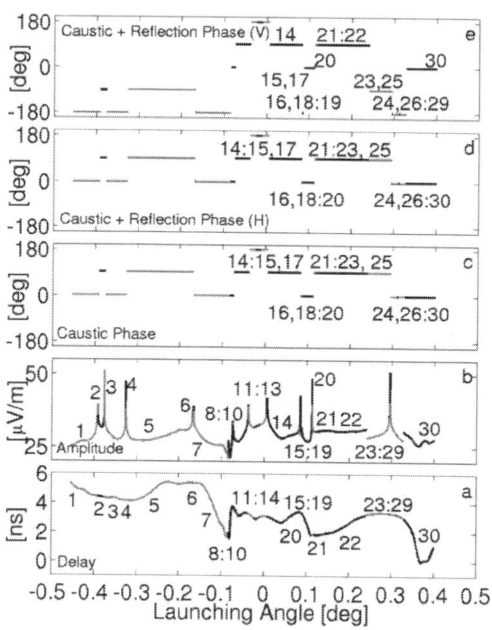

Figure 9. Green's function attributes for the field calculation: from bottom to top: delay(a), amplitude (b), caustic phase (c), caustic +reflection phase (H polarization), caustic + reflection phase (V polarization)

Each ray record refers is composed of the following parameters: the height of arrival, the travel time, which depend on the ray history; the amplitude and the phase, which depend on the ray history, on the reflection coefficients and on the system parameters.

The various quantities calculated by the ray tracing are referred to as Green's function attributes. Figure 9 shows among the attributes those that directly contribute to the field calculation.

3.2. Wave front interpolation

The field calculation along the vertical profile at the receiver range is performed through the interpolation of the different wavefronts contribution. Let K be the number of the wavefronts, each including N_k arrivals, the electric field associated to the ith arrival of the kth wavefront is:

$$\mathbf{E}(h_{ik}) = a_{ik} e^{j(2\pi f_0 T_{ik} + \phi_{ik})} \equiv \mathbf{E}_{ik} \tag{17}$$

where f is the frequency of operation of the link. The contribution of the kth wavefront to the total field is obtained through interpolation of the travel times, of the amplitudes and of the phases in the following way:

$$\mathbf{E}_k(h) = g_k(\mathbf{E}_{1k}, \mathbf{E}_{2k}, \ldots, \mathbf{E}_{N_k k}) \tag{18}$$

where g_k is a generic interpolating function and h is the height. As result, the interpolation of local values returns the field intensity in the target zone. Finally, the overall field is computed adding coherently all the wavefronts contributions:

$$\mathbf{E}(h) = \sum_{k=1}^{K} \mathbf{E}_k(h) \tag{19}$$

The vertical profile of the received power is obtained by applying locally the Poynting's theorem to a spherical surface with radius R:

$$P(h) = \frac{4\pi R^2}{\eta} \left| \mathbf{E}(h) \right|^2 \qquad (20)$$

being η the medium impedance and R the receiver range.

The solid line in the leftmost panel of figure 10 shows the power profile due to the direct arrival, while the dotted line represents the theoretical values predicted by Friis' formula (Balanis, 1996). Here, the vertical power profile is shown along the entire computational domain, from 0 up to 300 m, in order to show the effects of the tropospheric multipath on the received signal. The panel in the middle and the rightmost one of figure 10 show the power profile computed adding all the wavefront contribution at vertical (V) and horizontal (H) polarization.

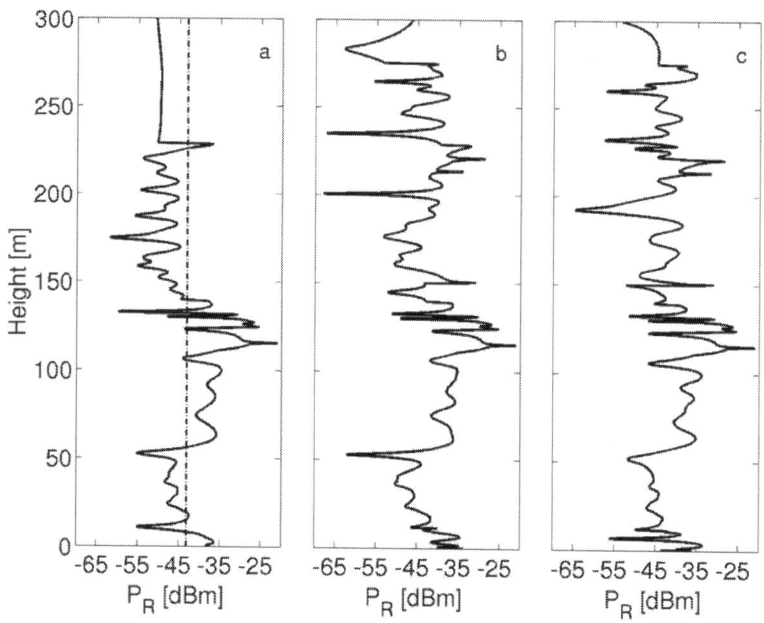

Figure 10. Received Power vertical profile, from left to right: direct arrival contribution (solid line) and theoretical arrive (Friis); direct + reflected contribution (vertical polarization); direct + reflected contribution (horizontal polarization)

4. RADIO CHANNEL ANALYSIS

4.1. Time domain approach

The aim of time domain approach is to estimate the radio channel response to a transmitted pulse with a given waveform $s(t)$. The use of the analytical extension concept let treat correctly the two distinct phase contributions of each arrival, i.e. the first one due to the propagation delay and the second one due to caustics and reflections, if any. The analytical extension of a base band signal $s(t)$ is given by:

$$\tilde{s}_{BB}(t) = s(t) + j\mathrm{H}\{\mathrm{s(t)}\} \tag{21}$$

where H represents the Hilbert transform operator. The corresponding pass band signal, which is transmitted on radio channel, is obtained with a frequency shift:

$$\tilde{s}_{RF} = \tilde{s}_{BB}(t)e^{j2\pi f_0 t} \tag{22}$$

The channel pulse response can be written as a sequence of ideal shifted pulses of the transmitted waveform with their own amplitude and phase:

$$w(t) = \sum_{k=1}^{K} A_K e^{j\phi_k} \delta(t - \tau_k) \tag{23}$$

where the amplitudes, the phases and the propagation delays have been calculated, respectively, through interpolation of a_{ik}, φ_{ik} and T_{ik} at height h. These procedure has low computational cost as the ray trajectories and propagation delay τ_k do not depend on the frequency and they can be computed once for all. In equation (23) some quantities such as amplitudes A_k and phases ϕ_k depends on the frequency of operation but using a narrow band approximation one can assume that amplitude and phase variations are negligible within the signal bandwidth. Under these hypothesis, the pass band received signal $\tilde{u}_{RF}(t)$ is given by the convolution between the pass band transmitted $\tilde{s}_{RF}(t)$ and the channel response (23):

$$\tilde{u}_{RF}(t) = \sum_{k=1}^{K} A_K e^{j\phi_k} \delta(t-\tau_k) \tilde{s}_{BB}(t-\tau_k) e^{j2\pi f_0 (t-\tau_k)}$$

(24)

The received base band signal $\tilde{u}_{BB}(t)$ is obtained multiplying the pass band signal $\tilde{u}_{RF}(t)$ by the exponential $e^{-j2\pi f_0 t}$:

$$\tilde{u}_{BB}(t) = \sum_{k=1}^{K} A_k e^{j\phi_k} \tilde{s}_{BB}(t-\tau_k) e^{-j2\pi f_0 \tau_k}$$

(25)

whose real part gives the received signal $u(t)$:

$$u(t) = \mathrm{Re}\{\tilde{u}_{BB}(t)\}$$

(26)

The panel in figure 11 reports the estimated field and the time delay as a function of height ranging between $(h_0 - \Delta h/2)$ and $(h_0 + \Delta h/2)$, being h_0= 125 m and Δh = 25 m observed at the receiver range. This represents the impulse response of the system computed at vertical polarization. The input signal is a Nyquist wavelet and it is drawn in the left image of figure 11 as a dotted line. Also, figure 11 shows the received signal (solid line) at four given height of the panel. Two in-phase but slightly delayed arrivals are evident in panel a). In panel b) there are 2 arrivals, with one of them with a π/4 phase shift, while in panel d) the arrivals are of opposite phase.

4.2. Frequency domain approach

The computation of the electric field in the target zone for a given frequency f_0 as reported in §3.2 strictly holds under the hypothesis of the transmission of a sinusoidal signal, which has an infinite duration and an infinitesimal band. As radio systems transmit signals with finite duration and band, there is the need to evaluate the response of the radio channel to the effects of frequency variation inside the system bandwidth of operation.

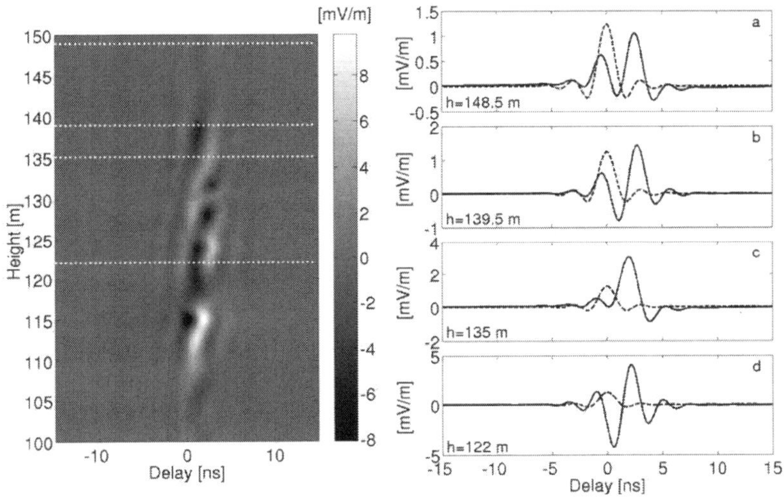

Figure 11. Impulse response panel (left). The image on the right shows the channel response to the transmitted Nyquist wavelet at four height values, sketched as a dotted white line in the panel on the left.

Let us assume that the radio link is transmitting a signal with bandwidth B around the carrier frequency f_0. Therefore, given the frequency f and using the equation (17) principle, the received electric field expression is:

$$E(f) = \sum_{k=1}^{K} A_k e^{j(2\pi f \tau_k + \phi_k)} \tag{27}$$

Applying equation (27) for each frequency in the interval spanning from $(f_0 - B/2)$ to $(f_0 + B/2)$ one obtains an estimate of the pass band radio channel transfer function. The evaluation of equation (27) at different frequencies requires to update the amplitude terms A_k and the phase terms

ϕ_k, which both depends (directly or indirectly) on the frequency of operation f.

Frequency domain approach can lead also to a characterization of radio channel in the time domain applying Fourier transform to the radio channel transfer function provided by equation (27). However, when the bandwidth of the transmitted signal is narrow, that is when $f_0 >> B$, this procedure leads to a waste of computational resources. A more efficient way to manage narrow band channels is to face the problem in the time domain.

The transfer function of the radio channel is shown in figure 12 together with the frequency behaviour computed at the same four height levels highlighted in the time domain analysis. Again the target zone is comprised between 100 and 150 m and the signal has vertical polarization. One can notices how the transfer function may change from a smooth behaviour, panel c) to a sharper one with a notch deepness of about 30 dB, panel a).

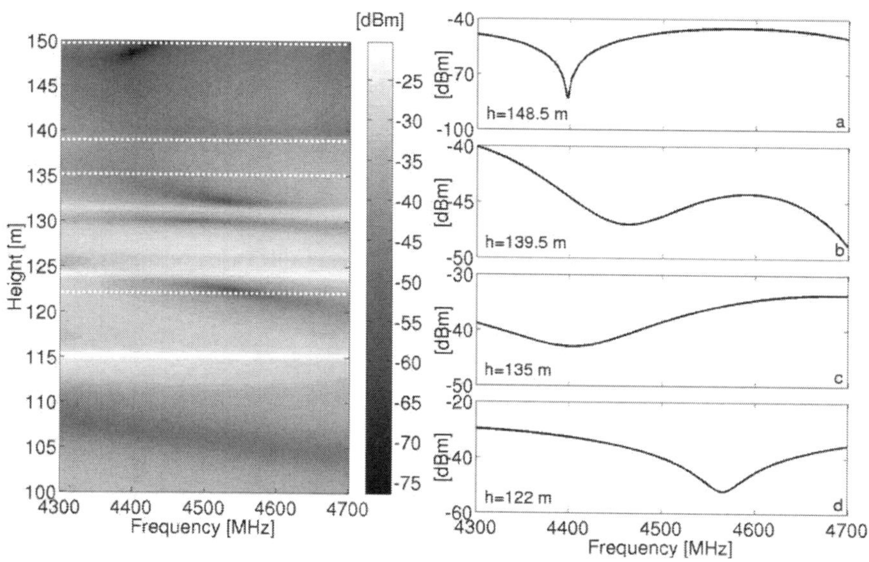

Figure 12. Transfer function panel (left). The image on the right shows the channel response to the transmitted Nyquist wavelet at four height values, sketched as a dotted white line in the panel on the left.

5. NUMERICAL VALIDATION

The ray tracing technique was developed to set up a prediction tool that is operative both as a MATLAB-based package and a C++ executable code. Unfortunately, beside an industrial application whose results are not available for publication at this stage, the computed field and power vertical profiles were not validated against measurements. Nevertheless the quality of the prediction performance can be observed when compared against canonical scenarios or other authors reference solutions.

Usually, when developing solver codes, one starts with the comparison of the predicted results with analytical or assessed solutions to check the correctness of the implementation. If this is the proper choice at that stage, it could lead to a flat and formal exposition of the methodology under study when describing the single steps towards the overall result. That is why the particular example chosen to illustrate the technique throughout §3 and §4 is far complicated, as it appears from the number and variety of wavefronts and from the aspects of the radio link characteristics both in time and in frequency domain. As the complexity could conceal errors or misevaluations, the author chose as reference solution an example in which a standard atmosphere model describes the velocity of the propagation medium over a flat perfect electric conductor ground. Even in such a simple example, the importance of the wave front discrimination could induce in errors when reconstructing the field.

5.1. Two ray example

The transmitter, in the origin of the abscissa axis, is at 150 m above the ground and the receiver range is 50 km apart. The atmospheric model is a standard exponential one with a gradient of 40 ppm in the first 1000m. Figure 13 shows the trajectories together with the two wavefront, one associated with the direct arrival and another with the reflected one. The ground is a perfect conductor with reflection coefficient equal to -1.

The two arrivals, direct and reflected, are shown in figure 14: on the leftmost panel the wavefront time delay dispersion results in 2 distinct contribution at the same instant. As in §3.1, the parameterization in the angular domain nicely unfolds the integration path (rightmost panel).

The predicted vertical received power profile is reported in figure 15, together with the direct arrival contributions given by Friis' formula. One

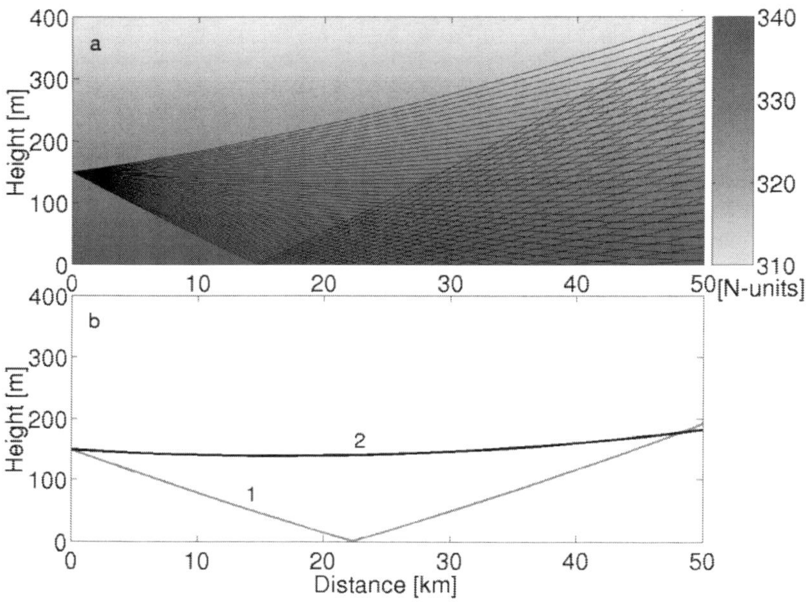

Figure 13. Ray pattern (top) and wavefronts (bottom)

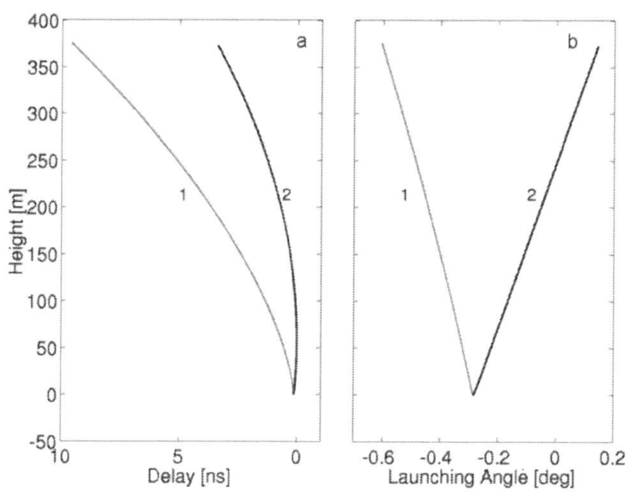

Figure 14. Wavefronts time delay dispersion (a) and launching angle distribution (b)

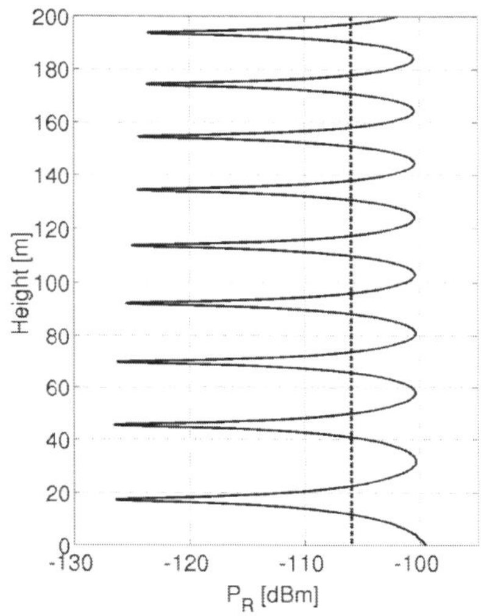

Figure 15. Received Power vertical profile (solid line) and direct arrival (dotted line)

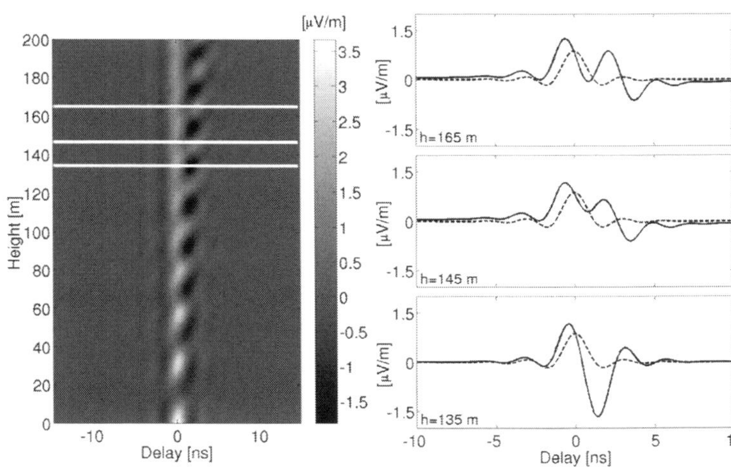

Figure 16. Impulse response panel (left). The image on the right shows the channel response to the transmitted Nyquist wavelet at three height values, sketched as a white line in the panel on the left.

can notice the typical pattern of zero and maxima of reception due to the rotating in-phase and counter-phase summation of the two arrivals.

Finally, figure 16 shows the impulse response panel calculated for the two ray model and vertical polarization. Again, the received signal shows different behaviour due to the atmospheric multipath as a function of the height.

6. CONCLUSION

The ray tracing technique based on the resolution of the Eikonal equation with Hamilton-Jacobi technique is well suited to model the dispersive effects of perturbed atmosphere on the performance of high-capacity digital radio channel.

The advantage of such an approach is the possibility of describing the medium with its variation in the lateral domain in presence of atmospheric multipath and of non uniform terrain profile.

The field profiling at the receiving site is obtained through 2,5D propagation as, notwithstanding all variations in the third dimension are taken equal to zero, the field amplitude is calculated correctly because of the paraxial approximation which takes place in a 3D domain.

At the time of writing the author is refining the code towards a full 3D implementation, which requires a reformulation in spherical coordinates, and the modelling of the mechanism of ground scattering and diffraction.

7. ACKNOWLEDGEMENTS

The author would like to thank Professor Giuseppe Drufuca (Politecnico di Milano) for his useful comments and suggestions and Simone Re (former research associate, Politecnico di Milano) for his contribution to the code development and calculations.

REFERENCES

1. M. Abramovitz, I. A. Stegun, 1970 Handbook of mathematical functions, Dover Publications, New York (USA).

2. Akbarpour R., Webster A.R., Ray-Tracing and parabolic Equation Methods in the Modeling of a Tropospheric Microwave Link, IEEE Transactions on Antennas and Propagation, 53 53 11 (November 2005), 27853791 , 0001-8926X.

3. Balanis C. A 1996 Antenna Theory: Analysis and Design, John Wiley & Sons, 0-47159-268-4 York (USA).

4. B. R. Bean, E.~. J. Dutton, 1968 Radio Meteorology, Dover Publications, New York (USA)

5. CCIR, Report 338 Propagation Data and Prediction Methods Required for Line-of-Sight Radio Relay Systems, Study Group 5 of the International Consultative Radio Committee (CCIR), Geneva (CH), 1978.

6. V. I. A. Červenŷ, I. Molotkov, Psencik, 1977 Ray method in Seismology, Univerzita Karlova Praha.

7. Driessen P.F., Prediction of multipath delay profiles in mountainous terrain, IEEE Journal on Selected Areas in Communications, 18 18 3 (March 2000), 336346 , 0733-8716

8. Du Castel 1966 Tropospheric Radiowave propagation beyond the Horizon, Pergamon Press, Oxford (UK).

9. Eresmaa R., Healy S., Järvinen H., Salonen K., Implementation of a ray-tracing operator for ground-based GPS Slant Delay observation modelling, Journal of Geophysical Research, 113 113 D11114, doi:10.1029/2007JD009256, (June 2008), 9 pages, 0148-0227

10. Farra V., Ray Tracing in Complex Media, Journal of Applied Geophysics, 30 30 1-2 , (April 1993), 5573 , 0926-9851

11. M. P. M. Hall, 1979 Effects of the troposphere on radio communication, IEEE Electromagnetic Waves Series, 0-90604-825-7 UK.

12. Keller J.B., Diffraction by an Aperture, Journal of Applied Physics, Vol. 28 4 (April 1957), 426444 .

13. Kravtsov Y.A., Orlov Y.I 1990 Geometrical Optics of Inhomogeneous Media, Springer-Verlag, 3-54051-944-0 York (USA).

14. S. N. Kružkov, solutions. Generalized, the. of, equations. Hamilton-Jacobi, Eikonal. of, I. type, Formulation of the problems; existence, uniqueness and stability theorems; some properties of the solutions, Mathematics of the USSR-Sbornik, 27 27 3 (1975), 406446 .

15. T. Kurner, D. J. Cichon, Concepts, results, 3 3D digital terrain-based wave propagation models: an overview, IEEE Journal on Selected Areas in Communication, 11 7 (September 1993), 10021012 , 0733-8716

16. M. Lebherz, W. Wiesbeck, W. Krank, A. Versatile, propagation. wave, for. model, V. H. F. U. H. F. the, considering. range, terrain. I. E. E. E. three-dimensional, on. Transactions, Antennas, Propagation, 40 40 10 October 1992), 11211131 , 0001-8926X.

17. Livingston D.C. 1970 The physics of microwave propagation, Prentice Hall, New Jersey (USA).

18. Operto M.S., Xu S., Lambaré G., Can we quantitatively image complex structures with rays?, Geophysics, Vol. 65 4 (July August 2000), 12231238 , 0016-8033

19. T. Pany, P. Pesec, G. Stangl, G. P. S. Atmospheric, Path. Slant, Delays, Tracing. Ray, Numerical. through, Models. a. Wather, Physics. Comparison, of. Chemistry, Earth. the, A. Part, Earth. Solid, Geodesy, 26 26 3 (2001), 183188 , 1474-7065

20. L. Sevgi, L. B. Felsen, A. New, For. Algorithm, Wave. Ground, Based. Propagation, A. On, Ray. Hybrid-Mode, International. Approach, of. Journal, Modelling. Numerical, 11 11 2 (March-April 1998), 87103 , Online 1099-1204

21. Y. Sun, Computation, 2 2D multiple arrival travel time fields by an interpolative shooting method, SEG Annual Meeting, Extended Abstracts (1992), 156159 , Tulsa (OK), USA.

22. Ulaby F.T. 1999 Fundamentals of applied electromagnetics, Prentice Hall, 0-13011-554-1 Jersey (USA).

23. Vinje V., Iversen E., Gjoystdal H., Traveltime and amplitude estimation using wavefront construction, Geophysics, Vol. 58 6 (August 1993), 11571166 , 0016-8033

9

CHAPTER

FIRST EVIDENCE OF SURFACE SH-WAVE PROPAGATION IN CUBIC PIEZOMAGNETICS

Aleksey Zakharenko

International Institute of Zakharenko Waves (IIZWs), Krasnoyarsk, Russia.

ABSTRACT

This theoretical work provides with results of characteristics calculation of the ultrasonic surface Zakharenko waves (USZWs) existing in piezomagnetic cubic monocrystals of class m3m that can be readily used for non-destructive testing. The piezomagnetic waves propagate in direction [101] corresponding to relatively easy magnetization for the following piezomagnetics: Galfenol, Terfenol-D, and CoFe2O4 with cubic structures. The phase velocities of the USZW-waves and the coefficient of magnetomechanical coupling (CMMC) K2 were calculated for the crystals. It was found that the coefficient K2 for piezomagnetics with Km2 > 1/3 and Km2 >> 1/3 is about 8% to 9%, where K2 = 2 (VUSZW,o – VUSZW,c)/VUSZW,o and Km2 = h2/(Cμ). Knowledge of piezomagnetic properties of cubic crystals makes possible the use of them in new products utilizing the phenomenon called the magnetoelectric effect. Also, this study is useful for possible application of cubic piezomagnetics in composite structures consisting of piezoelectric and (or) piezomagnetic materials and in the microwave technology. This broadens choice of possible piezomagnetic materials for utilization in various technical devices.

Keywords: piezomagnetic cubic monocrystals, Galfenol and Terfenol-D, ultrasonic surface Zakharenko waves

1. INTRODUCTION

Ferroic materials can display a spontaneous magnetization (ferromagnetic), polarization (ferroelectric) and strain (ferroelastic, shape memory alloy). Materials possessing two "ferro" properties simultaneously are called "biferroics". There are three kinds of biferroic materials: electroelastic, magnetoelastic, and magnetoelectric materials. The magnetoelectric materials (composites) can simultaneously possess ferromagnetic and ferroelectric properties. Composite materials or structures consisting of piezoelectric and piezomagnetic phases are able to facilitate the conversion of energy between electric and magnetic fields. Such phenomenon is called magnetoelectric (ME) effect. The ME effect of the composites was first reported by van Suchtelen [1]. The ME effect was then studied by van den Boomgaard [2], van Run et al. [3], and van den Boomgaard et al. [4] for BaTiO3/ CoFe2O4 composites. Possible applications of ME materials include magnetic-electric energy converting components; solid state non-volatile memory, multi-state memory, which can find application in quantum computing area; and electrical/optical polarization components, which can find applications in communication, light computing, and solid state memories based on spintronics [5,6]. Also, this research topic can be found in a recent review paper [7] by Fiebig. Note that the ME effect was originally predicted by P. Curie [8] in 1894.

Surface acoustic wave (SAW) devices are widely used in numerous branches of science and technology and their investigation especially in the case of interconnected physical fields is an important and actively developing branch of research and applications. About 120 years ago the first type of SAW was described by Lord Rayleigh [9] in connection with the problem of earthquakes. Bleustein [10] and Gulyaev [11] theoretically predicted that a pure shear-horizontal (SH) surface wave can be guided by the free surface of a piezoelectric half-space. Later, Maerfeld and Tournois

[12] investigated SH acoustic waves guided by the interface of two half-spaces, and Danicki [13] described the acoustic waves which can be guided by an embedded conducting plane in the electro-elastic materials of class 6 mm. It is thought that the simplest type of SH-SAW represents Love waves [14] in layered systems, consisting of an isotropic layer and an isotropic substrate. Recent developments in physics and technology made possible to construct new magneto-electro-elastic materials which demonstrate interconnection between magnetic, electricand elastic fields [15,16]. When the electric field was connected with the elastic one (piezoelectric materials) it brought up new and unexpected possibilities for science and technology. Connecting the magnetic field with the electric and elastic ones in magneto-electro-elastic materials suggests a range of new possibilities. Alshits et al. [17] conducted a qualitative investigation on the existence of surface waves in half-infinite anisotropic elastic media with piezoelectric, piezomagnetic, and ME effects. Soh and Liu [18] gave the existence conditions of interfacial SH-waves in a piezoelectric/piezomagnetic bi-material. Although much attention has been recently concentrated on magneto-electro-elastic materials and several dynamic problems have been solved by Hu and Li [19], Li [20], Chen, et al., [21], etc., the investigation of SAW propagation in transversely isotropic magneto-electro-elastic materials [22-25] of class 6 mm is currently an open actual subject. In addition to the transversely isotropic materials there are piezoelectric and piezomagnetic crystals with cubic symmetries, in which the surface BG-waves [10,11] cannot exist, according to [26] by Gulyaev and Hickernell. That is also true for the interfacial MTwaves [12]. However, new SH-SAWs called the ultrasonic surface Zakharenko waves (USZWs) were recently discovered in [27], which can propagate in cubic piezoelectrics. That is also true for cubic piezomagnetics. The purpose of this paper is to show the first evidence of the USZW existence in cubic piezomagnetics. It is possible to introduce piezomagnetic cubic crystals which are today highly-called for different technical devices.

The Cobalt Ferrite $CoFe_2O_4$ ($CoO{\cdot}Fe_2O_3$ or CFO) has the spinel ferrite structure with a stoichiometry denoted by AB_2O_4, where A is a divalent transition metal ion or a mixture of a trivalent and a monovalent metal ion. A spinel (see figure 1 and [28]) is a cubic structure in which O^{2-} ions form an fcc-lattice. It requires eight Formula equivalent AB_2O_4 to form a repeating unit cell which contains 32 O^{2-} ions, 64 tetragonal sites (A sites) and 32 octahedral sites (B sites). However, only 8 tetragonal and 16 octahedral sites are occupied so that the ratio of A-atoms/B-atoms is 1:2. CFO is known to have highly anisotropic magnetic properties and has been widely used in magnetic media and microwave device applications.

In 1971, the U.S. Navy developed an alloy of rare metals Tb and Dy with iron to create Terfenol-D (TbxDy1-xFey). The name "Terfenol-D" comes from a combination of TER for terbium, FE for iron, NOL for Naval Ordnance Laboratory, and D for dysprosium. Recently an iron/gallium alloy termed Galfenol has also been discovered at the Naval Surface Warfare Center (former NOL), where the magnetostrictive material Terfenol-D was discovered. It was found that magnetostriction in iron/gallium alloys peaks at a volume fraction of ~17% gallium. Unlike Terfenol-D representing a giant magnetostrictive material useable in practical operating conditions, FeGa-alloys are tough and not toxic, and can be machined and used without any special handling in devices where Terfenol-D may fracture. The cubic structure of Galfenol is shown in figure 2. Terfenol-D has been commercially available since the late 1980's and has since grown into an international industry. The Terfenol-D rods can be readily manufactured using the freestand-zone-melt process resulting in a cubic crystal structure [29] featuring a large magnetic anisotropy. Also, FeGa-alloys appear promising in filling the role as a mechanically robust material with substantial magnetostrictive capability [30]. Additional benefits of FeGa-alloys include substantially lower material costs than rare-earth based magnetostrictive alternatives. For example, the Terfenol-D (Tb27Dy73Fe195) raw material cost is approximately $ 0.50/g and current production techniques utilize carefully controlled crystal growth processes [31]. For comparison, the raw material cost for Fe81Ga19 is approximately $ 0.08/g [31]. Indeed,

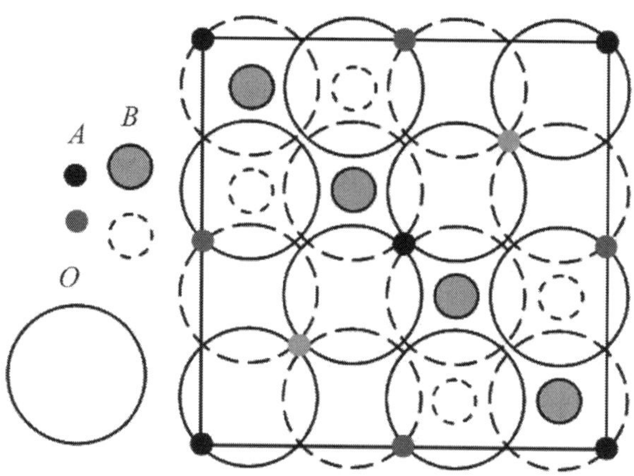

Figure 1. The normal spinel structure of Cobalt Ferrite CoFe2O4 (CoO·Fe2O3) with a stoichiometry denoted by AB2O4. Only the lower half of the unit cell is shown for projection on (001)

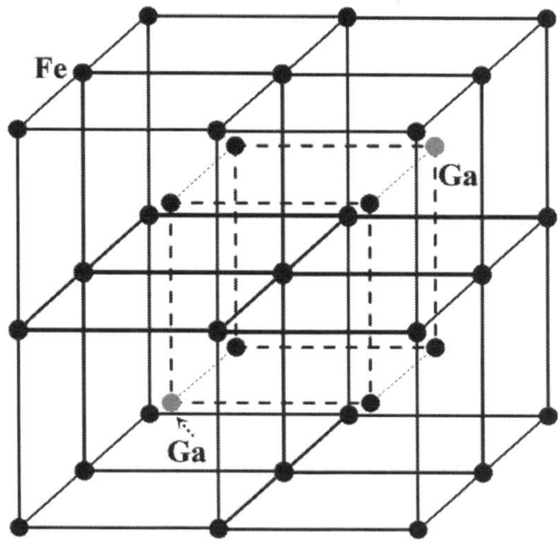

Figure 2. The bcc-structure of the Fe-17Ga alloy (α-Fe)

Terfenol-D can be used in magnetic transducers. It is wellknown that magnetic transducers are increasingly considered as actuators and sensors for numerous aerospace, aeronautic, automotive, industrial, and biomedical applications, for example see [32-34]. Recent progress in studying the cubic structure piezomagnetic crystals Galfenol and Terfenol-D of class m3m for utilization in transducers and sensors can be additionally found in [35-38]. Section 2 theoretically describes SAW propagation in piezomagnetic cubic crystals.

2. THEORY

The rectangular coordinate system (x1, x2, x3) is shown in figure 3. The suitable crystal cut and propagation direction was obtained by 450-rotation around the x2-axis directed along the lowest odd-order symmetry axis of a piezomagnetic cubic crystal. Therefore, the piezomagnetic waves with the anti-plane polarization propagate along the crystal surface in direction [101]. It is noted that the x2-axis is perpendicular to the sagittal plane. Theoretical description of wave propagation in cubic piezomagnetics is similar to that for cubic piezoelectrics, for example see [27,39,40]. The constitutive equations for a piezomagnetic material can be expressed in terms of the strains τ and the magnetic field H. Strains are related to mechanical displacements: τij = (∂Ui/∂xj + ∂Uj/∂xi)/2 [41]. The governing mechanical equilibrium is ∂σij/∂xj = 0 and the governing magnetostatic equilibrium is ∂Bi/∂xi = 0 where σij and Bi are the stress tensor and magnetic flux, respectively. A piezomagnetic medium possesses the elastic Cijkl and piezomagnetic hijk coefficients as well as the magnetic permeability coefficients μij and the medium density r.

It is necessary to write governing equations for the linear case. Constitutive relations are written as follows:

$$\sigma_{ij} = C^{H}_{ijkl}\tau_{kl} - h_{ijm}H_{m} \tag{1}$$

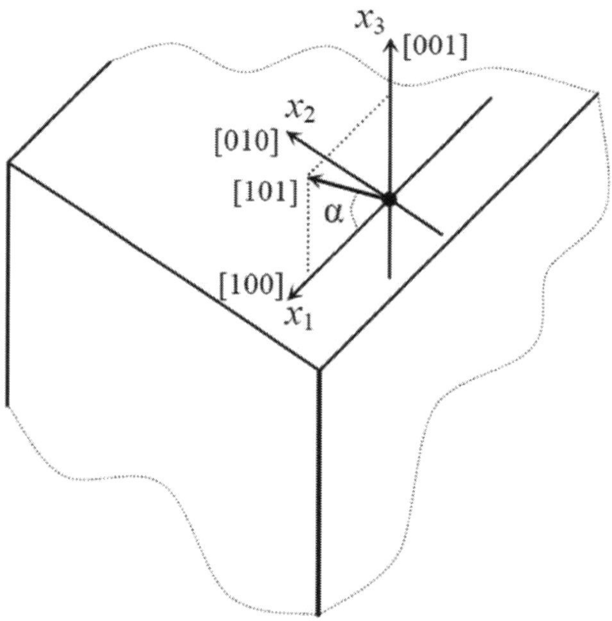

Figure 3. The rectangular coordinate system for a piezomagnetic cubic monocrystal, α = 45o

$$B_{m} = h_{mij}\tau_{ij} + \mu^{\tau}_{mn}H_{n} \tag{2}$$

in which σij and τij are the stress and strain tensors, respectively; Bm and Hm are components of the magnetic induction (i.e. magnetic flux) and magnetic field: Hm = – ∂ψ/∂xm, where ψ is the magnetic potential. The indices i, j, k, l, m, and n run from 1 to 3. According to Voigt's usual notation, Cijkl, hijm, and μmn can be written as 6 × 6, 3 × 6, and 3 × 3 matrices standing for the elastic, piezomagnetic, and magnetic tensors, respectively. They are thermodynamically defined as follows:

$$C_{ijkl}^{H} = \left(\frac{\partial \sigma_{ij}}{\partial \tau_{kl}} \right)_{H} \tag{3}$$

$$h_{mij} = h_{ijm} = -\left(\frac{\partial \sigma_{ij}}{\partial H_m} \right)_{\tau} = \left(\frac{\partial B_m}{\partial \tau_{ij}} \right)_{H} \tag{4}$$

$$\mu_{mn}^{\tau} = \left(\frac{\partial B_m}{\partial H_n} \right)_{\tau} \tag{5}$$

where the elastic constants Cijkl are defined at constant magnetic field and the magnetic constants μmn are defined at constant strain.

In the quasi-static approximation, the equations of motion of an elastic medium and magnetostatics are

$$\frac{\partial \sigma_{ij}}{\partial x_j} = \rho \frac{\partial^2 U_i}{\partial t^2} \quad \text{and} \quad \frac{\partial B_i}{\partial x_j} = 0 \tag{6}$$

In Equation (6), Ui denote the mechanical displacement components; t is time. Using Equations (1), (2), (6) and ψ = U4, one can write the coupled equations of motion for a piezomagnetic medium in the following form:

$$\rho \frac{\partial^2 U_i}{\partial t^2} = C_{ijkl} \frac{\partial^2 U_l}{\partial x_j \partial x_k} + h_{kij} \frac{\partial^2 \psi}{\partial x_j \partial x_k}$$
$$0 = h_{ijk} \frac{\partial^2 U_k}{\partial x_i \partial x_j} - \mu_{ij} \frac{\partial^2 \psi}{\partial x_i \partial x_j} \tag{7}$$

Solutions of homogeneous partial differential Equations (7) of the second order are found in the following plane wave form:

$U_i = U_i^0 \exp\left[j\left(\mathbf{K}_s\mathbf{R} - \omega t\right)\right]$ and $\psi = \psi^0 \exp\left[j\left(\mathbf{K}_s\mathbf{R} - \omega t\right)\right]$ where the index i runs from 1 to 3. Ui0 and ψ are initial amplitudes; j = (−1)1/2. ksr denotes the scalar multiplication of two vectors and ω is the angular frequency. {k1, k2, k3} = k{n1, n2, n3} are the components of the wavevector ks, {x1, x2, x3} are the components of the real space vector r, and {n1, n2, n3} are the directional cosines.

The coupled equations of motion can be readily written in the following simplified form, leaving only equations for waves with polarization perpendicular to the sagittal plane as well as non-zero components of the material tensors for the studied direction [101] of wave propagation:

$$\rho\frac{\partial^2 U_2}{\partial t^2} = C_{44}\left(\frac{\partial^2 U_2}{\partial x_1^2} + \frac{\partial^2 U_2}{\partial x_3^2}\right) - h_{16}\frac{\partial^2 \psi}{\partial x_1^2} + h_{34}\frac{\partial^2 \psi}{\partial x_3^2}$$

$$0 = -h_{16}\frac{\partial^2 U_2}{\partial x_1^2} + h_{34}\frac{\partial^2 U_2}{\partial x_3^2} - \mu_{11}\left(\frac{\partial^2 \psi}{\partial x_1^2} + \frac{\partial^2 \psi}{\partial x_3^2}\right) \tag{8}$$

In Equation (8), the mechanical displacement component U2 is directed along the x2-axis, see figure 3:

$$U_{2,4} = U_{2,4}^0 \exp\left[jk\left(n_1 x_1 + n_3 x_3 - V_{ph} t\right)\right] \tag{9}$$

where the phase velocity is defined as vph = w/k (k is the wavenumber in direction of wave propagation). For piezomagnetic cubic crystals, it is possible to write C44 = C66 = C and μ11 = μ33 = μ. It is noted that many propagation directions perpendicular to direction [010] can exist. It is also possible to cut a cubic crystal in order to study wave propagation in direction [101] with h14 = h36 = 0 and the non-zero piezomagnetic constants {−h16, h34} that is shown in Equation (8). The dependence of the normalized piezomagnetic constants on the cubic crystal cuts and propagation directions is shown in figure 4.

Substituting the mechanical displacement U2 and magnetic potential ψ = U4 of Equation (9) into Equation (8), the equations of motion can readily be written in the wellknown tensor form, using corresponding GL-components in the Green-Christoffel equation, (GLrw – δrwρVph) Ur = 0

[39,41]: $\mathrm{GL}_{22} = C(1+n_3^2)$, $\mathrm{GL}_{24} = \mathrm{GL}_{42} = -h(1-n_3^2)$ and $\mathrm{GL}_{44} = -\mu(1+n_3^2)$ with n3 = k3/k. In the GL-equation, r and w run from 1 to 4, δrw is the

Kronecker delta for r < 4 and w < 4, δ44 = 0, Ur = {U1, U2, U3, ψ}. Therefore, the following system of two homogeneous equations for pure SH-waves can be written as:

$$\begin{pmatrix} GL_{22} - C(V_{ph}/V_{t4})^2 & GL_{24} \\ GL_{42} & GL_{44} \end{pmatrix} \begin{pmatrix} U_2^0 \\ \psi^0 \end{pmatrix} = 0 \tag{10}$$

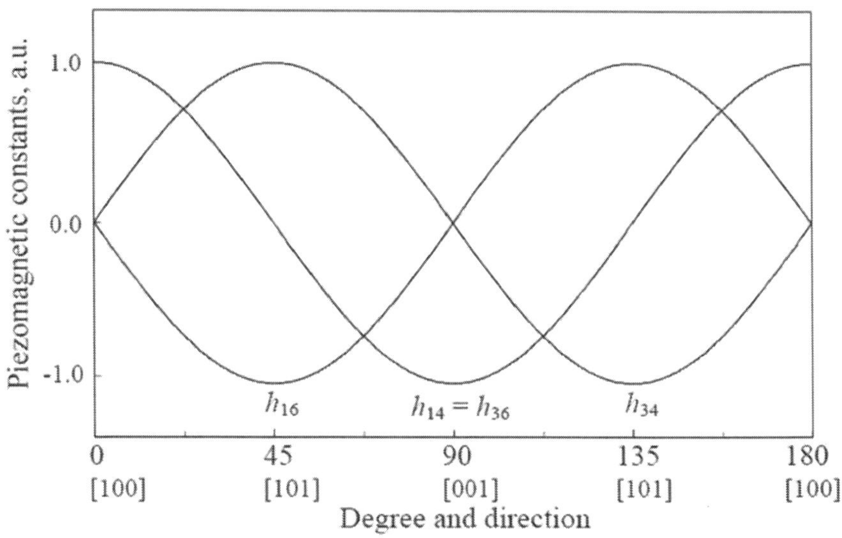

Figure 4. The dependence of the normalized piezomagnetic constants h14 = h36, h16, and h34 on the propagation directions with the Euler angles {0o, θ, 0o} for piezomagnetic cubic crystals, where there is rotation around the x2-axis

In Equation (10), the directional cosines are defined as follows: n1 ≡ 1, n2 ≡ 0 and n3 = n3. The velocity Vt4 is also defined as follows: Vt4 = (C44/ρ)1/2. Setting the matrix determinant equal to zero in Equation (10), the suitable phase velocity Vph satisfying boundary conditions discussed in the following section and four polynomial roots n3(p)(Vph), as well as the functions U20(Vph) and φ0(Vph), can then be found. For example, the functions can be taken in the following form: $\psi^0 = GL_{42}$ and $U_2^0 = -GL_{44}$.

It is noted that for the piezomagnetic surface Bleustein-Gulyaev waves, the GL24 and GL42 components are as follows: $\mathrm{GL}_{24} = \mathrm{GL}_{42} = h\left(1+n_3^2\right)$.

Substituting m = 1 + n32 and expanding the matrix determinant in Equation (10), the following polynomial can be obtained from Equation (10):

$$\left(1+K_m^2\right)m_3^2 - Dm_3 + 4K_m^2 = 0 \quad \text{with} \quad D = \left(\frac{V_{ph}}{V_{t4}}\right)^2 + 4K_m^2 \qquad (11)$$

of which two roots can be found as follows

$$m_3^{(1,2)} = \frac{D \pm \sqrt{D^2 - 16K_m^2\left(1+K_m^2\right)}}{2\left(1+K_m^2\right)} \qquad (12)$$

giving four polynomial roots of Equation (10)

$$n_3^{(1,2,3,4)} = \pm\sqrt{-1+m_3^{(1,2)}} \qquad (13)$$

The polynomial roots n3(1,2,3,4) actually represent the eigenvalues. For each eigenvalue n3(n), the corresponding eigenvector can be found in the following form: (U2(n), ψ(n)) where the index n runs from 1 to 4. The coefficient Km2 in Equations (11) and (12) is the static coefficient of the magnetomechanical coupling (CMMC):

$$K_m^2 = h_{16}^2 / \left(C_{44}\mu_{11}\right) \qquad (14)$$

It is also noted that the speed Vtm of the bulk SH-wave is given by the following Formula:

$$V_{tm} = V_{t4}\left(1+K_m^2\right)^{1/2} \qquad (15)$$

Analyzing the roots for propagation direction [101] in Equations (12) and (13), it can be found that all complex roots will be calculated when the expression under the square root in Equation (12) is negative. That fulfills for velocities Vph being lower than some velocity VKm obtained solving the following Equation from (12):

$$D^2 - 16K_m^2\left(1 + K_m^2\right) = 0 \tag{16}$$

and defined by the following formula:

$$V_{Km} = A_{Km}V_{t4} \text{ with } A_{Km} = 2\sqrt{K_m\sqrt{1+K_m^2} - K_m^2} \tag{17}$$

It is clearly seen in Equation (17) that the factor AKm shown in figure 5 is a function of the CMMC Km2. The other function f (Km2) = (1 + Km2)1/2 = Vtm/Vt4 from Equation (15) is also shown in figure 5. The function Akm (Km2 = K02 = 1/3) approaches the function f (Km2) = (1 + Km2)1/2 giving the following equality: VKm = Vtm (see figure 5). It is noted that only complex polynomial roots can exist when Vph < VKm. The CMMC K02 = 1/3 is readily found by substituting the velocity Vtm from Equation (15) instead of the phase velocity Vph in Equation (16). Note that when Km2 < K02 there are all imaginary roots for Vph > VKm, but a big Km2 > K02 gives real roots when Vph > VKm. It is also noted that only complex or imaginary roots with negative imaginary parts are chosen in order to cope with wave damping towards the depth of a crystal corresponding to negative values of the x3-axis shown in figure 3.

Table 1 lists experimental data for Galfenol (83% of Fe and 17% of Ga) obtained in [42,43]. It is clearly seen that the static CMMC Km2 for cubic Galfenol can be as high as~0.42 that is larger than 1/3, see also table 2. The Galfenol density ρ was taken from [31], which can depend on material synthesis. For example, [31] gives a Galfenol density of ~5900 kg/m3 for a typical sintered disk that is only 74% of maximum possible density. It is noted that the material density of Fe is ~7848 kg/m3 (see [44]). According to [31,42] as well as many Galfenol Workshops at the University of Maryland, the shear elastic constant c44 = c for Galfenol relatively slightly depends on both temperature and concentration of ga in fe. The elastic constant c can have values from ~120 GPa to ~135 gPa. According to [31], the static CMMC Km2 for cubic Terfenol-D can be usually about two times larger than that for Galfenol under the same conditions. However, it is interesting to compare some characteristics of Galfenol and Terfenol-D. Therefore, Terfenol-D with a very big Km2 ~1.2 was chosen from [45] for numerical calculations suggesting that Terfenol-d has the cubic symmetry. The piezomagnetics CoFe2O4 [16,46] is also listed in table 2 for comparison, suggesting the cubic symmetry for the crystal.

Since the experimental data listed in Table 1 is given for different experimental conditions, it is possible to calculate the single piezomagnetic constant h = h14 for cubic Galfenol, using known values of Km and C, as well as the measured values of magnetic permeability μ given in table 1. The strain τ and magnetic flux B can be also written based on the following linear constitutive piezomagnetic equations:

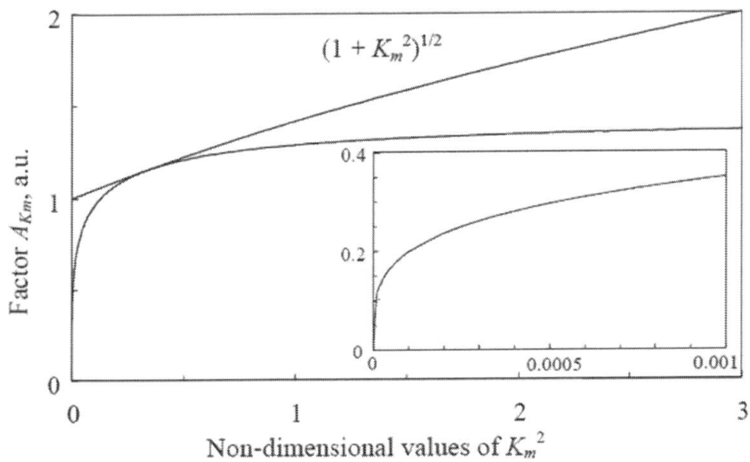

Figure 5. The dependence of both the factor AKm and the function (1 + Km2)1/2 on the values of Km2 from 0 up to 3. The insertion shows the AKm(Km2) for small values of the Km2

Table 1. The material characteristics following [42,43] for Galfenol Fe0.83Ga0.17 (m3m); EB = 7.7 × 1010 [N/m2]. The magnetic permeability coefficient of a vacuum is μ0 = 4π × 10−7 [H/m]~12.566371 × 10−7 [(V × s)/(A × m)]. The static coefficient of magnetomechanical coupling (CMMC) Km in the last column was also announced in [42,43]

Stress Ξ, 10^6 [N/m^2]	q, 10^{-9} [m/A]	μ^σ/μ_0,	K_m
16.0	58	285	0.65
27.5	34	174	0.54
50.5	22	108	0.47
85.0	16	70	0.42

Table 2. The elastic C44 and piezomagnetic h14 constants as well as the material density ρ and magnetic permeability coefficient μ11 for the piezomagnetic cubic crystals: Galfenol [42,43], CoFe2O4 [16,46], and Terfenol-D [45]. The coefficient of magnetomechanical coupling (CMMC) Km2 in the last column can be calculated with Formula (14)

Material (symmetry group)	Density, ρ [kg/m^3]	C_{44}, 10^{10} [N/m^2]	h_{14} [N/(A×m)]	$\mu_{11} = \mu^{\varepsilon}$, 10^{-6} [N/A^2]	K_m^2
Galfenol (m3m)	7973	12.70	3331.34	206.83	0.4225
Galfenol (m3m)	7973	12.70	2395.05	154.90	0.2916
Galfenol (m3m)	7973	12.70	1722.32	105.74	0.2209
Galfenol (m3m)	7973	12.70	1273.98	72.45	0.1764
$CoFe_2O_4$ (m3m)	5300	4.53	550.0	157.00	0.0425
Terfenol-D (m3m)	9250	0.60	167.7	3.97	1.1807

$$\tau = S^H \sigma + qH \tag{18}$$

$$B = q^* \sigma + \mu^\sigma H \tag{19}$$

where sH is the compliance at constant magnetic field H and μσ is the magnetic permeability at a constant stress σ. Note that Equation (19) incorporates the direct magnetostrictive effect in which magnetic flux is generated by stresses in the piezomagnetic material. The piezomagnetic constants q and q* are thermodynamically defined as follows, assuming q = q*:

$$q = \left(\partial\tau/\partial H\right)_\sigma \text{ and } q^* = \left(\partial B/\partial\sigma\right)_H \tag{20}$$

The piezomagnetic coefficient q symbolizes the axial magnetostriction coefficient and represents the change in strain per change in magnetic field at a constant stress. The parameter q* represents the change in magnetic induction per change in stress at a constant magnetic field.

The short circuit elastic modulus EB represents the stiffest-material condition, which occurs when all available magnetic energy has been transduced into elastic potential energy. When energy is transferred from the elastic to the magnetic regime the effective modulus decreases to the value of EH. It has been shown by invoking linear transduction considerations that these two elastic moduli are related by the magnetomechanical coupling coefficient Km2 (0 ≤ Km2 ≤ 1) as follows [47]:

$$E^H = E^B\left(1 - K_m^2\right) \tag{21}$$

In a similar manner, the intrinsic or uncoupled magnetic permeability μσ of Equation (19) can be reduced to a value corresponding to the constant strain permeability με because of the energy conversion from the magnetic to the elastic regime. The permeabilities are related by:

$$\mu^\varepsilon = \mu^\sigma\left(1 - K_m^2\right) \tag{22}$$

It is emphasized that expressions (21) and (22) are only valid for linear systems and significant corrections are necessary in order to extend these formulae to the full nonlinear regime. However, they are highly useful to illustrate the exchange mechanisms taking place in magnetostrictive transduction.

The static CMMC Km2 can be also written as follows:

$$K_m^2 = q^2 / \left(S^H \mu^\sigma\right) \tag{23}$$

Therefore, the single non-zero piezomagnetic constant h can be calculated with the following formula, using Equations (14), (22), and (23):

$$h = K_m \sqrt{C\mu^\sigma\left(1 - K_m^2\right)} \tag{24}$$

Table 2 lists the piezomagnetic constant h for cubic Galfenol corresponding to the experimentally measured values of the magnetic permeability μσ at constant stress σ.

For the free space, Laplace's equation of type Dψ = 0 is written in the following form: (k12 + k32) ψ0 = 0 where μ0 is the magnetic constant for the free space. The magnetic potential for the free space can be written as follows: ψ0 = F(0)exp (− k1x3) exp [j (k1x1 − wt)]. The potential ψ0 must decrease according to any increase in the coordinate x3 (see figure 3). It is now possible to use the material constants in mechanical and magnetic boundary conditions.

3. BOUNDARY CONDITIONS FOR SURFACE SH-WAVES

The mechanical boundary conditions of a magnetoelastic material which occupies the half-space x3 < 0 must be satisfied together with the magnetic conditions of a magnetically closed (B3 = 0) or magnetically open (ψ = 0) surface. The realization of each of these boundary conditions is described in [17]. There is the single mechanical boundary condition for the normal component of the stress tensor: σ32 = 0 at x3 = 0, where,

$$\sigma_{32} = \sum_{p=1,2} F^{(p)}\left[Ck_3^{(p)}U_2^{(p)} + hk_3^{(p)}U_4^{(p)}\right] \tag{25}$$

Also, there are the magnetic boundary conditions: continuity of the normal component B3 of the magnetic flux at x3 = 0 being the interface between a vacuum (B3f) and the crystal surface (B3 = B3f) where,

$$B_3 = \sum_{p=1,2} F^{(p)}\left[hk_3^{(p)}U_2^{(p)} - \mu k_3^{(p)}U_4^{(p)}\right]$$
$$B_3^f = -F^{(0)}\psi_0^f\,\mathrm{j}k_1\mu_0 \tag{26}$$

and continuity of the magnetic potential U4 = ψ at x3 = 0 (ψ = ψ f) where

$$\psi = \sum_{p=1,2} F^{(p)}\psi^{(p)} \quad \text{and} \quad \psi^f = F^{(0)}\psi_0^f \tag{27}$$

Therefore, using Equations (25)-(27), two homogeneous equations for the case of a magnetically closed surface can be readily written as follows:

$$\begin{pmatrix} Ck_3^{(1)}U_2^{(1)} + hk_3^{(1)}U_4^{(1)} & Ck_3^{(2)}U_2^{(2)} + hk_3^{(2)}U_4^{(2)} \\ \mathrm{j}\mu_0 k_1 U_4^{(1)} & \mathrm{j}\mu_0 k_1 U_4^{(2)} \end{pmatrix}\begin{pmatrix} F^{(1)} \\ F^{(2)} \end{pmatrix} = 0 \tag{28}$$

For a magnetically open surface, the following homogeneous equations can also be written in the corresponding form:

$$\begin{pmatrix} Ck_3^{(1)}U_2^{(1)} + hk_3^{(1)}U_4^{(1)} & Ck_3^{(2)}U_2^{(2)} + hk_3^{(2)}U_4^{(2)} \\ hk_3^{(1)}U_2^{(1)} - (\mu k_3^{(1)} - \mathrm{j}\mu_0 k_1)U_4^{(1)} & hk_3^{(2)}U_2^{(2)} - (\mu k_3^{(2)} - \mathrm{j}\mu_0 k_1)U_4^{(2)} \end{pmatrix} \begin{pmatrix} F^{(1)} \\ F^{(2)} \end{pmatrix} = 0 \tag{29}$$

Therefore, the corresponding boundary-condition determinants (BCDs) of matrices in Equations (28) and (29) for both cases represent complex numbers.

The complete mechanical displacement U2Σ and magnetic potential ψΣ = U4Σ can be written in the plane wave form as follows:

$$U_{2,4}^{\Sigma} = \sum_{p=1,2} F^{(p)} U_{2,4}^{0(p)} \exp\left[\mathrm{j}k\left(n_1 x_1 + n_3^{(p)} x_3 - V_{ph} t\right)\right] \tag{30}$$

The corresponding weight functions F(1) and F(2) are found from Equations (28) and (29), which can give the same eigenvectors (U20(1), ψ0(1)) and (U20(2), ψ0(2)) for two equal eigenvalues n3(1) = n3(2), and hence $F^{(1)} = -F^{(2)}$. It is obvious that for this case the weight factors $F^{(1)} = -F^{(2)}$ will zero the complete mechanical displacement U2Σ and magnetic potential ψΣ in Equations (28) and (29) giving "latent" characteristics in Equation (30). On the other hand, unequal eigenvalues n3(1) and n3(2) give different eigenvectors (U20(1), ψ0(1)) and (U20(2), ψ0(2)). It is thought that the Vph-solutions with the "latent" characteristics must be experimentally verified due to the fact that some elements of crystals symmetry (screw axis or glide reflection) can be broken near the surface. The same relates to simple reflections and axis if the surface restricts the crystal in such a way that it breaks some simple elements of the crystal class group. It is also noted that Vph- solutions cannot be found when the surface BleusteinGulyaev waves are studied that is an additional difference for finding the BG-waves and USZW-waves.

4. RESULTS AND DISCUSSIONS

Acoustic wave propagation along the surface of piezoelectric material in direction [101] as an instability problem can sustain the surface SH-waves in cubic piezoelectrics that was recently discovered in [27]. Concerning piezomagnetics, it is thought that piezomagnetic SH-SAW can also propagate in direction [101] in cubic piezomagnetics, for instance, crystals of class m3m. Indeed, all cubic crystals of the class possess no piezoelectricity according to [48,49]. Note that concerning transversely-isotropic materials, solutions for the surface BG-waves in piezomagnetics can be written in an identical manner to the wave solutions in piezoelectrics [50]. It is also noted that the velocity VBGmo of piezomagnetic surface BG-wave propagating along the magnetically open surface of transversely-isotropic piezomagnetics (treating the simplest case) can be found with the following well-known formula:

$$V_{BGmo} = V_{tm}\left(1-\left(\frac{K_m^2}{\left(1+K_m^2\right)\left(1+\mu/\mu_0\right)}\right)^2\right)^{1/2} \tag{31}$$

The velocity VBGmc of piezomagnetic BG-wave with the magnetically closed surface can be calculated with the following formula:

$$V_{BGmc} = V_{tm}\left(1-\left(\frac{K_m^2}{1+K_m^2}\right)^2\right)^{1/2} \tag{32}$$

Note that Formula (31) for the velocity VBGmo is not applicable for the case of cubic piezomagnetics. For example, formula (31) gives VBGmo~4760.093893 m/s for Galfenol with the largest value of Km2 > 1/3 in the first row in table 3. This value of VBGmo is close to that of Vtm, but not to the value of VKm (see table 3) representing the right result for cubic piezomagnetics with Km2 > 1/3. The same there is for Terfenol-D with Km2~1 in table 2, for which the incorrect velocity calculated with Formula (31) is VBGmo~1179.199962 m/s.

For cubic piezomagnetics with Km2 < 1/3, Formula (31) also gives values of VBGmo being closer to corresponding values of Vtm than true values of VUSZW,o. For example, the velocity VUSZW,o for Galfenol in the fourth row in table 3 with the smallest value of Km2 < 1/3 (see also table 2) is VUSZW,o~4328.763803 m/s. However, Formula (31) gives VBGmo ~4328.783915 m/s that is closer to Vtm~ 4328.798060 m/s. Therefore, the differences ΔBG = VBGmo – Vtm and ΔUSZW = VUSZW,o – Vtm are ~0.014 m/s and ~0.034 m/s, respectively. For comparison, for the piezomagnetic CoFe2O4 with the smallest value of Km2 ~4% from table 2, the true velocity VUSZW,o in table 3 is already very close to the velocity VBGmo ~2985.08072 m/s, which is also very close to the velocity Vtm. Note that the relative magnetic permeability μ/μ0 in Formula (31) for piezomagnetics is about one order larger than the corresponding relative dielectric constant ε/ε0 for piezoelectrics for the same Km2 = Ke2, where Ke2 is the static coefficient of electromechanical coupling (CEMC), and ε and ε0 are the single electric constants for a piezoelectrics and free space, respectively. That can result in mm/s-difference for piezomagnetics compared with cm/s-difference for piezoelectrics between the corresponding BG-wave velocity and bulk wave velocity Vt. That can result in significantly larger penetration depth of surface BG and USZW-waves for piezomagnetics towards negative values of the x3-axis in figure 3. It is also noted that the velocity VBGmc for a magnetically closed surface can be readily calculated with Formula (32), with which the velocity of interfacial Maerfeld-Tournois waves [12] is also calculated. The interfacial Maerfeld-Tournois waves can propagate along the interface of two transverselyisotropic half-spaces of class 6 mm. Concerning cubic piezomagnetics, it is expected that some interfacial waves can also propagate that can be further researched.

Table 3. The velocities vKm, Vt4, and Vtm (all in m/s) as well as the velocities VUSZW,o and VUSZW,c of the ultrasonic surface Zakharenko waves for [101] propagation direction in the piezomagnetic cubic crystals. The last column gives values of the K2 calculated with Formula (33). All the Galfenols correspond to those listed in tables 1 and 2

Material (symmetry group)	Structure	V_{Km}	V_{t4}	V_{tm}	$V_{USZW,o}$	$V_{USZW,c}$	K^2, %
Galfenol (m3m)	bcc	4740.796256	3991.085010	4760.101550	4740.750567	4545.299858	8.24
Galfenol (m3m)	bcc	4530.187138	3991.085010	4535.795584	4535.727732	4418.693612	5.16
Galfenol (m3m)	bcc	4360.495418	3991.085010	4409.913035	4409.878451	4337.133082	3.30
Galfenol (m3m)	bcc	4217.255080	3991.085010	4328.798060	4328.763803	4279.858496	2.26
$CoFe_2O_4$ (m3m)	Spinel	2396.906402	2923.554311	2985.080877	2985.080691	2982.595557	0.17
Terfenol-D (m3m)	-	1048.738635	805.3872662	1189.319685	1047.759141	999.920958	9.13

Using the velocities VUSZW,o and VUSZW,c for the USZWwaves, it is possible to evaluate the coefficient of magnetomechanical coupling (CMMC) K2 for the piezomagnetic cubic crystals listed in table 3. Indeed, it is thought that it is possible to use the well-known formula that is used for piezoelectrics. Therefore, the coefficient K2 can be evaluated with the following formula:

$$K^2 = 2\left(\frac{V_{USZW,o} - V_{USZW,c}}{V_{USZW,o}}\right) \tag{33}$$

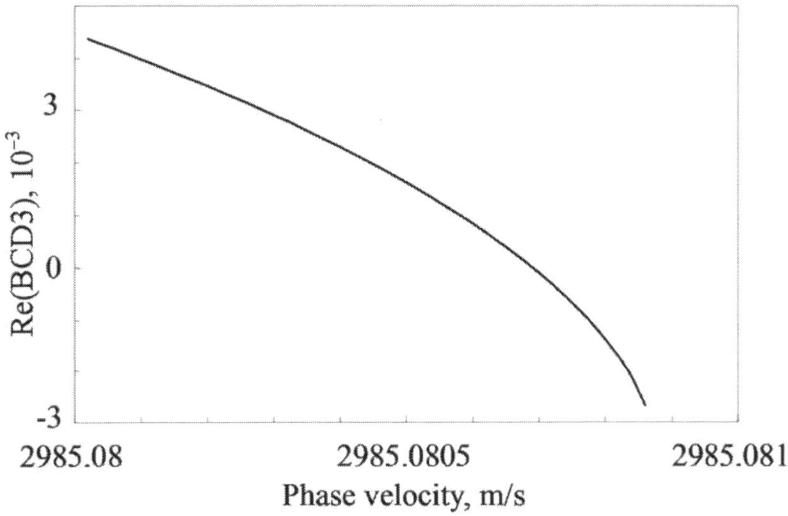

Figure 6. The behavior of the boundary-condition determinant (BCD3) for the piezomagnetic monocrystal CoFe2O4

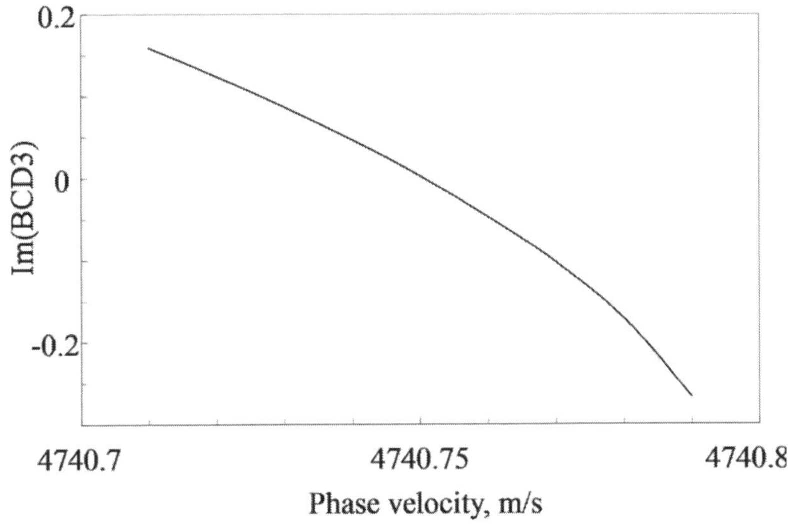

Figure 7. The behavior of the boundary-condition determinant (BCD3) for the piezomagnetic bcc-cubic monocrystal of the Fe-17Ga alloy for Km = 0.65

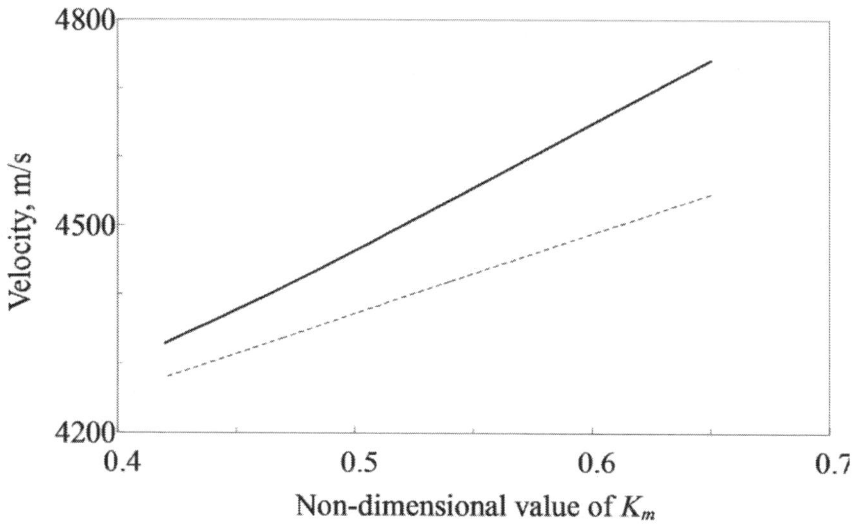

Figure 8. The dependence of the phase velocity Vph on the coefficient Km for the Fe-17Ga alloy, where the functions VUSZW,o(km) and VUSZW,c(km) are shown by solid and dashed lines, respectively

For Galfenol in tables 2 and 3, the evaluated value of K2 is as high as 2.26% for the smallest value of Km2. The value of K2 increases with increase in Km2 to the biggest value of 8.25%. This value of 8.25% for the piezomagnetic Galfenol in the first row of table 3 can be compared with that for Terfenol-D in the Table: K2(Terfenol-D)~ 9.13%. It is clearly seen in the second and third tables that a big value of Km2~1.2 for Terfenol-D being about three times larger than the biggest value of Km2 for Galfenol does not give a significant rise in the coefficient K2. The rise is smaller than 1%. That can be explained using the following fact: the velocity VUSZW,o for cubic piezomagnetics with a giant coefficient Km2 is situated slightly below the speed VKm, but not slightly below the speed Vtm (see also figure 5). That results in lower coefficient K2. Therefore, it can be concluded that the coefficient K2 for cubic piezomagnetics with a giant Km2 is about 8% to 9% and cannot be larger than 10%. It is thought that the coefficient K2 for the surface BG-waves in transversely-isotropic piezomagnetics with a giant Km2 can be larger than 10%, according to Formulae (31)–(33) and figure 5. The sample behaviors of the boundarycondition determinant (BCD3) around the phase velocity solutions for the piezomagnetic monocrystals CoFe2O4 and Galfenol are shown in figures 6 and 7, respectively. Figure 8 shows the dependence of the Vph on the coefficient Km for the Fe-17Ga alloy. Note that any change in the material density will change both the velocities Vt4 and Vtm (Vt4), but not the coefficients Km2 and K2.

5. CONCLUSIONS

This paper addresses the theoretical investigations of piezomagnetic acoustic waves propagating along the surface of cubic crystals. The ultrasonic surface Zakharenko waves (USZWs) can propagate in cubic piezomagnetics when direction [101] is parallel to the crystal surface. The computer simulation was done for the widelycalled piezomagnetic crystals Galfenol, Terfenol-D, and

CoFe2O4 suggesting the cubic symmetry. The cubic piezomagnetics Galfenol and Terfenol-D can possess large values of Km2, where Km2 is the static coefficient of magnetomechanical coupling. The simulations have shown that for a cubic crystal with a large Km2 > 1/3 (even Km2 > 1 for Terfenol-D) the coefficient K2 is as big as 8%-9% and not larger than 10%. It is noted that a big Km2 > 1 does not give a significant rise to the coefficient K2 for the case of Km2~1/3. That is due to the fact that velocity solutions for large values of Km2 > 1/3 are found just below the velocity VKm, but not just below the velocity Vtm. This theoretical study of cubic piezomagnetics allows use of suitable surfaces of piezomagnetic cubic crystals in non-destructive testing and evaluation that was also recently done for cubic piezoelectrics.

6. ACKNOWLEDGEMENTS

The author is grateful to all the Referees for their constructive comments.

REFERENCES

1. J. V. Suchtelen, "Product Properties: A New Application of Composite Materials," Philips Research Reports, Vol. 27, No. 1, 1972, pp. 28-37.

2. J. V. D. Boomgaard, D. R. Terrell, R. A .J. Born and H. F. J. I. Giller, "An in Situ Grown Eutectic Magnetoelectric Composite-Material: 1.

Composition and Unidirectional Solidification," Journal of Materials Science, Vol. 9, No. 10, 1974, pp. 1705-1709.

3. M. J. G. V. Run, D. R. Terrell and J. H. Scholing, "An in Situ Grown Eutectic Magnetoelectric Composite-Material: 2. Physical Properties," Journal of Materials Science, Vol. 9, No. 10, 1974, pp. 1710-1714.

4. J. V. D. Boomgaard, A. M. J. G. V. Run and J. V. Suchtelen, "Piezoelectric-Piezomagnetic Composites with Magnetoelectric Effect," Ferroelectrics, Vol. 14, 1976, pp. 727-728.

5. V. E. Wood and A. E. Austin, "Magnetoelectric Interaction Phenomena in Crystals," In A. J. Freeman, H. Schmid, Eds., Gordon and Breach Science Publishers, Newark, New York, 1975, pp. 181-194.

6. Y. N. Venevtsev, V. V. Gagulin and I. D. Zhitomirsky, "Material Science Aspects of Seignette-Magnetism Problem," Ferroelectrics, Vol. 73, 1987, pp. 221-248.

7. M. Fiebig, "Revival of the Magnetoelectric Effect," Journal of Physics D: Applied Physics, Vol. 38, No. 8, 2005, pp. 123-152.

8. P. Curie, "Sur la Symétrie des Phénomènes Physiques: Symétrie d'un Champ électrique et d'un Champ Magnétique," Journal de Physique, 3e Série, Vol. 3, 1894, pp. 393-415.

9. L. Rayleigh, "On Waves Propagated along the Plane Surface of Elastic Solids," Proceedings of the London Mathematical Society, Oxford, Vol. 17, 1885, pp. 4-11.

10. J. L. Bleustein, "A New Surface Wave in Piezoelectric Materials," Applied Physics Letters, Vol. 13, No. 12, 1986, pp. 412-413.

11. Y. V. Gulyaev, "Electroacoustic Surface Waves in Solids," Soviet Physics Journal of Experimental and Theoretical Physics Letters, Vol. 9, 1969, pp. 37-38.

12. C. Maerfeld and P. Tournois, "Pure Shear Elastic Surface Wave Guide by the Interface of two Semi-Infinite Media," Applied Physics Letters, Vol. 19, No. 4, 1971, pp. 117-118.

13. E. Danicki, "New Interfacial Shear Wave in Piezoelectrics," Applied Physics Letters, Vol. 64, No. 8, 1994, pp. 969-970.

14. E. H. Love, "Some Problems of Geodynamics," Cambridge University Press, London, 1911.

15. W. Nan, "Magnetoelectric Effect in Composites of Piezoelectric and Piezomagnetic Phases," Physical Review B, Vol. 50, No. 9, 1994, pp. 6082-6088.

16. S. Srinivas, J. Y. Li, Y. C. Zhou and A. K. Soh, "The Effective Magnetoelectroelastic Moduli of Matrix-Based Multiferroic Composites," Journal of Applied Physics, Vol. 99, No. 4, 2006, pp. (043905)1-7.

17. V. I. Al'shits and A. N. Darinskii and J. Lothe, "On the Existence of Surface Waves in Half-Infinite Anisotropic Elastic Media with Piezoelectric and Piezomagnetic Properties," Wave Motion, Vol. 16, No. 3, 1992, pp. 265- 283.

18. A.-K. Soh and J.-X. Liu, "Interfacial Shear Horizontal Waves in a Piezoelectric-Piezomagnetic Bi-Material," Philosophical Magazine Letters, Vol. 86, No. 1, 2006, pp. 31-35.

19. K.-Q. Hu and G.-Q. Li, "Constant Moving Crack in a Magnetoelectroelastic Material under Anti-Plane Shear Loading," International Journal of Solids and Structures, Vol. 42, No. 9-10, 2005, pp. 2823-2835.

20. X.-F. Li, "Dynamic Analysis of a Cracked MagnetoelectrOelastic Medium under Anti-Plane Mechanical and InPlane Electric and Magnetic Impacts," International Journal of Solids and Structures, Vol. 42, No. 11-12, 2005, pp. 3185-3205.

21. J. Chen, E. Pan and H. Chen, "Wave Propagation in Magneto-Electro-Elastic Multilayered Plates," International Journal of Solids and Structures, Vol. 44, No. 3-4, 2007, pp. 1073-1085.

22. K. Mori and M. Wuttig, "Magnetoelectric Coupling in Terfenol-D/polyvinylidenedifluoride Composites," Applied Physics Letters, Vol. 81, No. 1, 2002, pp. 100-101.

23. M. I. Bichurin, D. A. Filippov, V. M. Petrov, V. M. Laletsin, N. Paddubnaya and G. Srinivasan, "Resonance Magnetoelectric Effects in Layered Magnetostritive-Piezoelectric Composites," Physical Review B, Vol. 68, No. 13, 2003, pp. (132408)1-4.

24. K. Srinivas, G. Prasad, T. Bhimasankaram and S. V. Suryanarayana, "Electromechanical Coefficients of Magnetoelectric PZT-CoFe2O4 Composite," Modern Physics Letters B, Vol. 14, No. 17-18, 2000, pp. 663-674.

25. P. Yang, K. Zhao, Y. Yin, J.G. Wan and J.S. Zhu, "Magnetoelectric Effect in Magnetostrictive/Piezoelectric Laminate Composite Terfenol-D/LiNbO3 [(zxtw)-129°/ 30°]," Applied Physics Letters, Vol. 88, No. 17, 2006, pp. (172903)1-3.

26. Y. V. Gulyaev and F. S. Hickernell, "Acoustoelectronics: History, Present State and New Ideas for a New Era," Acoustical Physics, Vol. 51, No. 1, 2005, pp. 81-88.

27. A. Zakharenko, "New Solutions of Shear Waves in Piezoelectric Cubic Crystals," Journal of Zhejiang University SCIENCE, Vol. 8, No. 4, 2007, pp. 669-674.

28. W. Kleber, "An Introduction to Crystallography," VEB Verlag Technik, Berlin, 1970.

29. M. J. Dapino, "Nonlinear and Hysteretic MagnetomechaNical Model for Magnetostrictive Transducers," PhD Thesis, Iowa State University, 1999.

30. E. Clark, J. B. Restorff, M. Wun-Fogle, T. A. Lograsso and D. L. Schlagel, "Magnetostrictive Properties of Bodycentered Cubic Fe-Ga and Fe-Ga-Al Alloys," IEEE Transactions on Magnetics, Vol. 36, No. 5, 2000, pp. 3238- 3240.

31. R. A. Kellog, "Development and Modeling of Iron-Gallium Alloys," PhD Thesis, Iowa State University, 2003.

32. M. J. Dapino, "On Magnetostrictive Materials and their Use in Adaptive Structures," International Journal of Structural Engineering and Mechanics, Vol. 17, No. 3-4, 2004, pp. 303-329.

33. M. J. Dapino, F. T. Calkins and A. B. Flatau, "Magnetostrictive Devices," Wiley Encyclopedia of Electrical and Electronics Engineering, In. J. G. Webster, Ed., John Wiley and Sons Inc., New York, Vol. 12, 1999, pp. 278- 305.

34. J. B. Restorff, "Magnetostrictive Materials and Devices," Encyclopedia of Applied Physics, Vol. 9, 1994, pp. 229- 244.

35. R. Chung, R. Weber and D. Jiles, "A Terfenol-D Based Magnetostrictive Diode Laser Magnetometer," IEEE Transactions on Magnetics, Vol. 27, No. 6, 1991, pp. 5358- 5243.

36. M. J. Dapino, R. C. Smith and A. B. Flatau, "StructuralMagnetic Strain Model for Magnetostrictive Transducers," IEEE Transactions on Magnetics, Vol. 36, No. 3, 2000, pp. 545-556.

37. P. R. Downey and A. B. Flatau, "Magnetoelastic Bending of Galfenol for Sensor Applications," Journal of Applied Physics, Vol. 97, No. 10, 2005, pp. (10R505)1-3.

38. J. Atulasimha, A. B. Flatau and R. A. Kellogg, "Sensing Behavior of Varied Stoichiometry Single Crystal Fe-Ga," Journal of Intelligent Material Systems and Structures, Vol. 17, No. 2, 2006, pp. 97-105.

39. G. W. Farnell and E. L. Adler, "Elastic Wave Propagation in Thin Layers," Physical Acoustics: Principles and Methods, In. W. P. Mason and R. N. Thurston Eds., Academic Press, New York, Vol. 9, 1972, pp. 35-127.

40. C. Lardat, C. Maerfeld and P. Tournois, "Theory and Performance of Acoustical Dispersive Surface Wave Delay Lines," Proceedings of the IEEE, Vol. 59, No. 3, 1971, pp. 355-368.

41. V. E. Lyamov, "Polarization Effects and Interaction Anisotropy of Acoustic Waves in Crystals," MSU Publishing, Moscow, 1983.

42. E. Clark, K. B. Hathaway, M. Wun-Fogle, J. B. Restorff, T. A. Lograsso, V. M. Keppens, G. Petculescu and R. A. Taylor, "Extraordinary Magnetoelasticity and Lattice Softening in Bcc Fe-Ga Alloys," Journal of Applied Physics, Vol. 93, No. 10, 2003, pp. 8621-8623.

43. E. Clark, "Structural Fe-based Alloys with High Magnetostriction," Galfenol Workshop, University of Maryland, 29 January 2004.

44. A. Zakharenko, "On Cubic Crystal Anisotropy for Waves with Rayleigh-Wave Polarization," Non-Destructive Testing and Evaluation, Vol. 21, No. 2, 2006, pp. 61-77.

45. T. J. C. Liu and C. H. Chue, "On the Singularities in a Bimaterial Magneto-Electro-Elastic Composite Wedge under Anti-Plane Deformation," Composite Structures, Vol. 72, No. 2, 2006, pp. 254-265.

46. F. Ramirez, P. R. Heyliger and E. Pan, "Free Vibration Response of Two-Dimensional Magneto-Electro-Elastic Laminated Plates," Journal of Sound and Vibration, Vol. 292, No. 3-5, 2006, pp. 626-644.

47. E. Clark, "Magnetostrictive Rare Earth-Fe2 Compounds," Ferromagnetic Materials, In. E. P. Wohlfarth, Ed., North-Holland Publishing Company, Amsterdam, Vol. 1, Chapter 7, 1980, pp. 531-589.

48. V. I. Al'shits and V. N. Lyubimov, "On Existence of Piezoelectric and Piezomagnetic Properties of Anisotropic Media," Soviet Kristallographia, Moscow, Vol. 35, No. 2, 1990, pp. 483-484.

49. D. Royer and E. Dieulesaint, "Elastic Waves in Solids I," Springer-Verlag, New York, 2000.

50. Y. V. Gulyaev, I. E. Dikshtein and V. G. Shavrov, "Magnetoacoustic Surface Waves in Magnetic Crystals at Reorientation Phase Transitions," Uspekhi Phizicheskikh Nauk, Moscow, Vol. 167, No. 7, 1997, pp. 735-750.

10

CHAPTER

PROPAGATION IN TUNNELS: EXPERIMENTAL INVESTIGATIONS AND CHANNEL MODELING IN A WIDE FREQUENCY BAND FOR MIMO APPLICATIONS

J.-M.Molina-Garcia-Pardo,[1] M. Lienard,2 and P. Degauque[2]

(1) Departamento de Tecnología de la Información y la Comunicación, Technical University of Cartagena

(2) Télécommunications, Interférences et Compatibilité Electromagnétique (TELICE), Institut d'Electronique, Microélectronique et Nanotechnologie (IEMN), University of Lille

ABSTRACT

The analysis of the electromagnetic field statistics in an arched tunnel is presented. The investigation is based on experimental data obtained during extensive measurement campaigns in a frequency band extending from 2.8 GHz up to 5 GHz and for a range varying between 50 m and 500 m. Simple channel models that can be used for simulating MIMO links are also proposed.

1. INTRODUCTION

Narrowband wireless communications in confined environments, such as tunnels, have been widely studied for years, and a lot of experimental results have been presented in the literature in environmental categories ranging from mine galleries and underground old quarries to road and railway tunnels [1–4].

However, in most cases, measurements dealt with channel characterization for few discrete frequencies, often around 900 MHz and 1800 MHz. For example, in [5, 6] Zhang et al. report statistical narrowband and wideband measurement results. In [7], results on planning of the Global System for Mobile Communication for Railway (GMS-R) are presented. In [8], simulations and measurements are also described in the same GSM frequency band. In [9], the prediction of received power in the out-of-zone of a dedicated short range communications (DSRC) system operating inside a typical arched highway tunnel is discussed, and in this case the channel impulse response was measured with a sounder at 5.2 GHz whose bandwidth is on the order of 100 MHz. Recently, in [10], measurement campaigns have been performed in underground mines in the 2–5 GHz band but the results cannot be extrapolated to road and railway tunnels since the topology is quite different. In a mine gallery, roughness is very important, the typical width is 3 m, the geometry of the cross-section is not well defined and lastly, there are often many changes in the tunnel direction.

Furthermore, to increase the channel capacity in tunnels, space diversity both at the mobile and at the fixed base station can be introduced. However, good performances of multiple input multiple output (MIMO) techniques can be obtained under the condition of a small correlation between paths relating each transmitting and receiving antennas. This decorrelation is usually ensured by the multiple reflections on randomly distributed obstacles, giving often rise to a wide spread in the direction of arrival of the rays. On the contrary, a tunnel plays the role of an oversized waveguide and decorrelation can be due to the superposition of the numerous hybrid modes supported by the structure [11]. Experimental results at 900 MHz for a $(4, 4)$MIMO configuration, are described in [12]. This paper shows that the antenna arrays must be put in the transverse plane of the tunnel to minimize the coupling between elements.

The objective of this work is thus to extend the previous approaches by investigating the statistics of the electric field distribution in the 2.8–5 GHz frequency range in a tunnel environment for MIMO applications. Empirical formulas based on the experimental results are also proposed.

We proceed in two steps: (1) determination of the mean path loss and of the statistical distribution of the average field which can be received by the various antennas of an MIMO system. This first approach can thus be used to determine the average power related to the **H** matrix of an MIMO link, (2) field distribution and correlation in a transverse plane.

The paper is distributed as follows. Section 2 explains the experiments in detail and more specifically the environment and methodology of the measurements that has been followed. Section 3 investigates path loss and axial correlation while, in Section 4, field statistics in the transverse plane are analyzed. Section 5 deals with the transverse spatial correlation and Section 6 presents the principle of modeling the MIMO channel and gives an example of application. Finally, Section 7 summarizes the contributions of the present work and gives conclusions.

2. ENVIRONMENT, MEASUREMENT EQUIPMENT, AND METHODOLOGY

2.1. Description of the Environment

The measurement campaign was performed in a 2-way tunnel, situated in the French Massif Central mountains. This straight tunnel, 3.4 km long, has a semicircular shape, as shown in Figure 1. The diameter of the cylindrical part is 8.6 m and the maximum height of the tunnel is 6.1 m. The tunnel was empty with no pipes, cables, or lights. However, every 100 m there are small safety zones, 1 m wide and few meters long, where an extinguisher is hung. It is difficult to estimate the roughness accurately but it is on the order of a centimetre. The tunnel was closed to traffic during the experiments, to make measurements in stationary conditions.

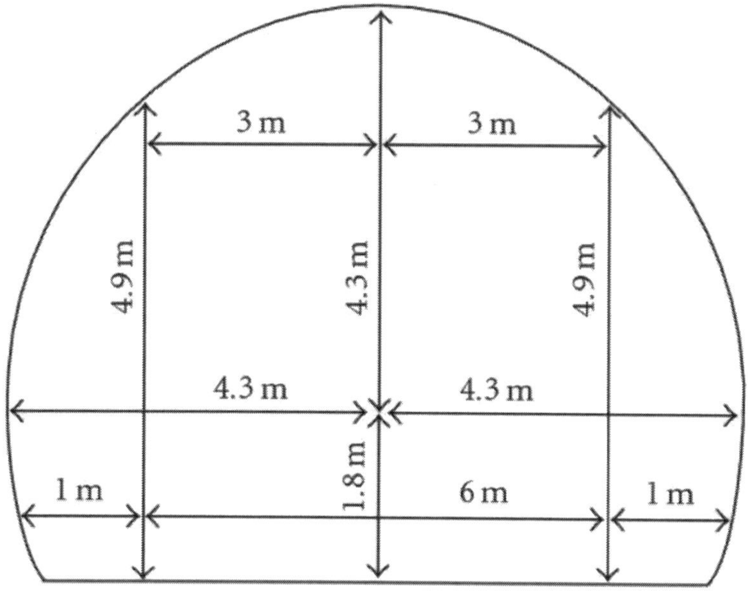

Figure 1. Cross-section of the tunnel.

2.2. Measurement Equipment

Since we want to explore the channel response in a very wide frequency band (2.8–5 GHz), we have chosen to make measurements in the frequency domain rather than in the time domain, so as to get better accurate results. The complex channel transfer function between the transmitting (Tx) and receiving (Rx) antennas has thus been obtained by measuring the S_{21} parameter with a vector network analyzer (VNA Agilent E5071B). The Rx antenna is directly connected to one port of the VNA using a low attenuation coaxial cable, 4 m long, a 30 dB low-noise amplifier being inserted or not, depending on the received power. Using a coaxial cable to connect the Tx antenna to the other port of the VNA would lead to prohibitive attenuation, the maximum distance between Tx and Rx being 500 m. The signal of the Tx port of the VNA is thus converted to an optical signal which is sent through fibre optics, converted back to radio frequency and amplified. The signal feeding the vertical biconical transmitting (Tx) antenna has a power of 1 W. The phase stability of the fibre optics link has been checked and the calibration of the VNA takes amplifiers, cables, and optic coupler into account. The block diagram of the channel sounder is depicted in Figure 2.

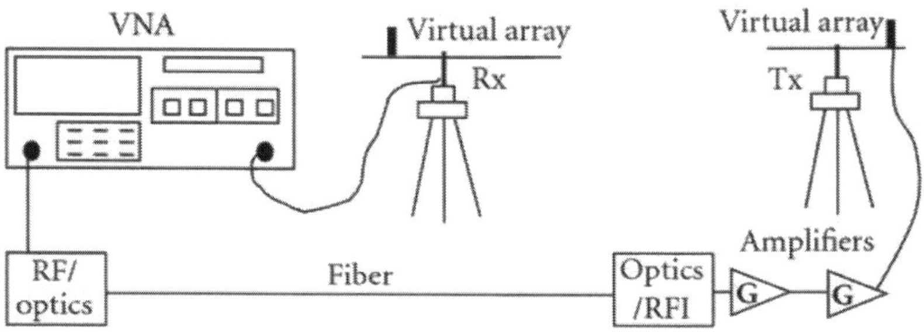

Figure 2. PRINCIPLE OF THE CHANNEL SOUNDER SETUP.

The wideband biconical antennas (Electrometrics EM-6116) used in this experiment have nearly a flat gain, between 2 and 10 GHz. Indeed, the frequency response of the two antennas has been measured in an anechoic chamber, and the variation of the antenna gain was found to be less than 2 dB in our frequency range. Nevertheless, we have subtracted the antenna effect in the measurements, as it will be explained in Section 3.

It must also be emphasized that, in general, the radiation pattern of wideband antennas is also frequency dependent. This is not a critical point in our case since, in a tunnel, only waves impinging the tunnel walls with a grazing angle of incidence contribute to the total received power significantly. This means that, whatever the frequency, the angular spread of the received rays remains much smaller than the 3 dB beam width of the main antenna lobe in the E plane, equal to about 80°, the antenna being nearly omnidirectional in the H plane.

Since the channel transfer function may also strongly depend on the position of the antennas in the transverse plane of the tunnel, both Tx and Rx antennas were mounted on rails. The position mechanical systems are remote controlled, optic fibres connecting the step by step motors to the control unit.

2.3. Methodology

The channel frequency response has been measured for 1601 frequency points, equally spaced between 2.8 and 5 GHz, leading to a frequency step of 1.37 MHz.

The rails supporting the Tx and Rx antennas were put at a height of 1 m and centred on the same lane of this 2-lane tunnel. For each successive axial distance *d*, both the Tx and Rx antennas were moved in the transverse plane on a distance of 33 cm, with a spatial step of 3 cm, corresponding to half a wavelength at 5 GHz. A $(12, 12)$ transfer matrix is thus obtained, the configuration of the measurements being schematically described in Figrue 3. Fine spatial sampling was chosen for measurements in the transverse plane because, as recalled in the introduction, antenna arrays for MIMO applications have to be put in this plane to minimize correlation between array elements.

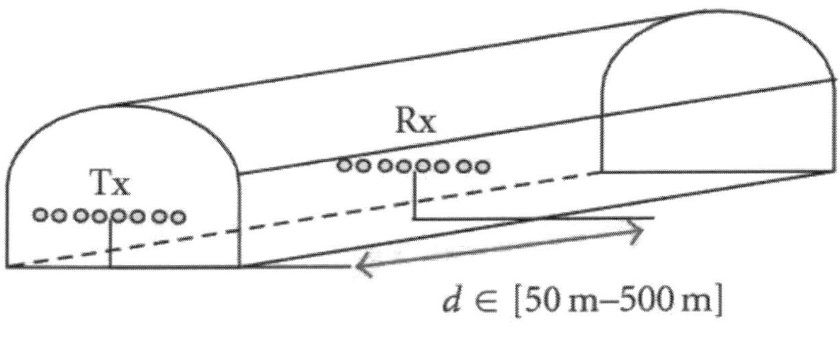

Figure 3. CONFIGURATION OF THE WIDEBAND MIMO MEASUREMENTS.

Due to the limited time available for such an experiment and to operational constraints, it was not possible to extensively repeat such measurements for very small steps along the tunnel axis. In the experiments described in this paper, the axial step was chosen equal to 4 m when 50 m $<d<$ 202 m and to 6 m when 202 m $<d<$ 500 m. This is not critical because we are interested, in the axial direction, by the mean path loss and by the large-scale fluctuation of the average power received in the transverse plane. At each Tx and Rx position, 5 successive recordings of field variation versus frequency are stored and averaged.

It must be noted that in the case of a single input single output (SISO) link, a number of papers have already been published on the small-scale variation of a narrowband signal along the tunnel axis. For example, [13] describes results of experiments carried out in a wide tunnel at a frequency of 900 MHz. A summary of the measurement parameters and equipment characteristics is summarized in Table 1.

Table 1. Equipment characteristics and measurement parameters.

Frequency band	2.8–5 GHz
Number of frequency points	1601
Antenna	Biconical antenna (Electrometrics EM-6116)
Transmitter power	20 dBm
Dynamic range	>100 dB
Position in the transverse plane	12 positions every 3 cm ($\lambda/2$ at 5 GHz)
Positions along the longitudinal axis	From 50 m to 202 m every 4 m From 202 m to 500 m every 6 m
Number of acquisitions at each position	5

3. PATH LOSS AND CORRELATION ALONG THE LONGITUDINAL AXIS

3.1. Path Loss

The path loss is deduced from the measurement of the $S_{21}(f,d)$scattering parameter. However, as briefly mentioned in the previous section, it can be more interesting to subtract the effects of the variation of the antenna characteristics with frequency by introducing a correction factor $C(f)$. We have thus made preliminary measurements by putting the two biconical antennas, 1 m apart, in an anechoic room. Let $S_{21}^{anech}(f)$ be the scattering parameter measured in this configuration. The correction factor is thus given by $C(f) = |S_{21}^{anech}(f)| - \langle |S_{21}^{anech}(f)| \rangle$, where $\langle x \rangle$ means the average of x over the frequency band.

The path loss in tunnel, taking this correction into account, is given by

$$\mathrm{PL}(f,d) = -20 \cdot \log_{10}(|S_{21}(f,d)|) - (-20 \cdot \log_{10} C(f)). \tag{1}$$

Figure 4 shows the variation of $\mathrm{PL}(f,d)$ versus frequency, for $d = 50$ m. The fluctuation of the field amplitude is due to the combination in phase or out of phase of the various modes excited by the transmitting antenna, the phase of the propagation constant depending on frequency but also on the order of the hybrid modes propagating in the tunnel. To extract the variation of the mean path loss versus frequency, it is interesting to average such curves, obtained at any distance d, for the various transverse positions of the antennas. Furthermore, one can also average over few frequencies, considering a frequency bandwidth smaller than the channel coherence bandwidth. In this example, the coherence bandwidth being on the order of 10 MHz, $\mathrm{PL}(f,d)$ was averaged over 7 frequencies around f, the frequency step being 1.37 MHz, and over the 144 successive combinations of the transverse positions of the Tx and Rx antennas. The average value $\langle \mathrm{PL}(f,d) \rangle$ is also plotted in Figure 4.

The curves "measurements" in Figure 5 represent the variation of $\langle \mathrm{PL}(f,d) \rangle$ versus axial distance at 3 and 5 GHz. The path loss, at 3 GHz, corresponding to free-space conditions, has been also plotted. We see that, in this frequency range, the path loss is only slightly dependent on frequency.

To deduce from these curves a simple theoretical model of the mean path loss $\overline{\mathrm{PL}}(f,d)$, these curves must be smoothed again by introducing a running mean over the axial distance. To get a very simple approximate analytical expression of $\overline{\mathrm{PL}}(f,d)$, it is assumed that $\overline{\mathrm{PL}}(f,d)$ is the product of two functions, one depending on f and one depending on d [14].

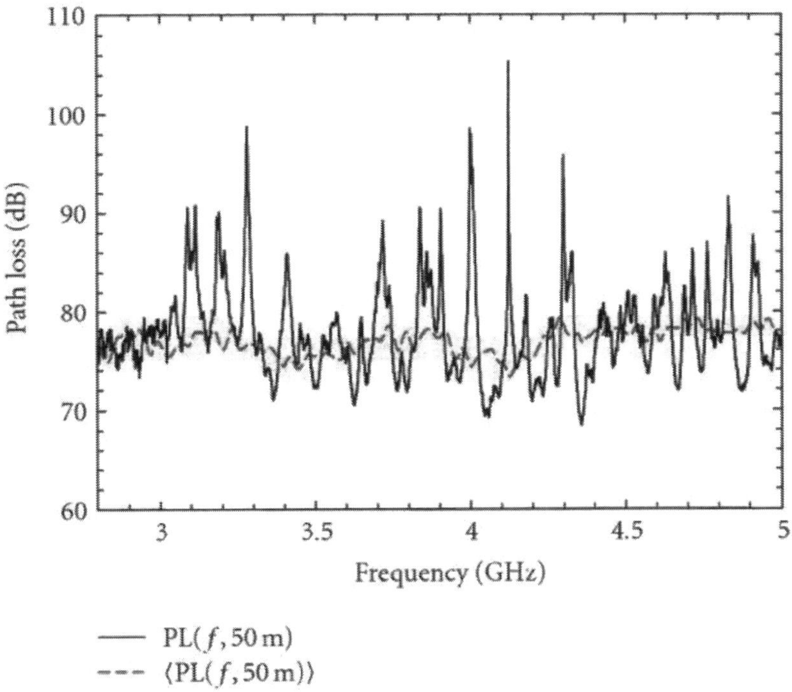

Figure 4. PATH LOSS $PL(f,d)$ between two antennas for $d = 50$ m and path loss $\langle PL(f,d)\rangle$ averaged over the transverse positions of the antennas and over 7 frequencies in a 10 MHz band.

Furthermore, it is usually expressed in terms of two path loss exponents, $n_{PL_0\_f}$ and $n_{PL_0\_d}$ which indicate the rate at which the path loss decreases with frequency and distance, respectively, [15]. This leads to

$$\overline{PL}(f,d) = (PL_0 + 10 \cdot n_{PL_0\_f} \log_{10}(f(\text{GHz}))) + 10 n_{PL_0\_d} \log_{10}(d). \tag{2}$$

The constant PL_0 and the path loss exponents have been determined by minimizing the mean square error between the measurements and the model. The following values were found: $PL_0 = 86$ dB, $n_{PL_0\_f} = 0.82$, and

$n_{PL_0\_d} = 0.57$. The corresponding curves for 3 and 5 GHz have also been plotted in Figure 5. It must be outlined that all these values were deduced from measurements between 50 and 500 m and consequently, they are valid only in this range of axial distance.

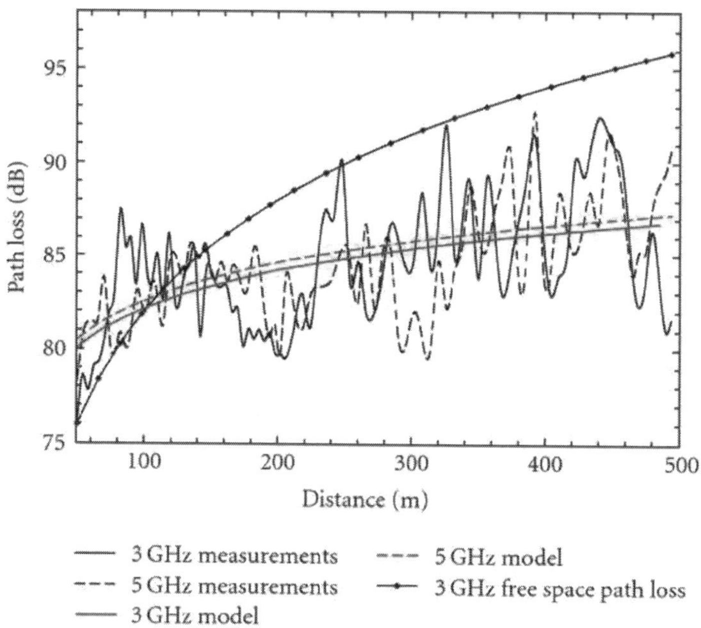

Figure 5. AVERAGE PATH LOSS (CURVES "MEASUREMENTS") AND MEAN PATH LOSS DEDUCED FROM THE MODEL (CURVES "MODEL") AT 3 AND 5 GHZ.

It can be interesting to compare this value of $n_{PL_0\_d}$ to those already published in the literature and corresponding to attenuation factors measured for ultra-wideband systems in indoor environments. However, in this case, the range is much smaller, typically below 50 m. In line of sight (LOS) conditions, values from 1.3 to 1.7 were reported by [16, 17], while for non-LOS, $n_{PL_0\_d}$ may reach 2 to 4 as mentioned in [18, 19]. The small value that we have obtained comes from the guiding effect of the tunnel.

The comparison between $\langle \mathrm{PL}(f,d)\rangle$ and the predicted path loss $\overline{\mathrm{PL}}(f,d)$ shows that the difference in their values is characterized by a standard deviation $\sigma_{\mathrm{PL}} = 2.7$ dB. $\langle \mathrm{PL}(f,d)\rangle$ can thus be modeled by (2) and by adding a random variable $X_{\sigma_{\mathrm{PL}}}$ with zero mean and standard deviation σ_{PL}:

$$\langle \mathrm{PL}(f,d)\rangle_{\mathrm{model}} = \overline{\mathrm{PL}}(f,d) + X_{\sigma_{\mathrm{PL}}}. \tag{3}$$

3.2. Axial Correlation

One can expect that the variation of the average received power between one transverse plane and another will depend on the distance *d*, high-order propagating modes suffering important attenuation at large distances. To study this point, we have calculated, for a given frequency, the amplitude ρ_{axial} of the complex correlation coefficient between the $(12,12)$ transfer matrix elements measured at a distance *d* and the matrix elements measured at the distance $d+\Delta d$, *d* varying between 50 m and 500 m. Let us recall that the step Δd is equal 4 m while 50 m $< d <$ 202 m and 6 m when 202 m $< d <$ 500 m.

Curves in Figure 6 give the variation of ρ_{axial} for three frequencies: 3, 4, and 5 GHz. As one can expect from the modal theory, the correlation is an increasing function of distance. At 3 GHz, for example, the correlation between 2 receiving arrays, 4 m apart, varies from 0.6 at 50 m, to reach an average value of 0.9 at a distance of 200 m. If we now compare results obtained at 3 and 5 GHz, we see that the correlation increases less rapidly at 5 GHz, because high-order modes suffer less attenuation. By examining the shape of these curves, we observe two regions: the first one, at short distance from the transmitter, where the correlation increases nearly linearly, and the other where the average value of the correlation does not vary appreciably. A two-slope model seems thus well suited to fit the average variation of the correlation function.

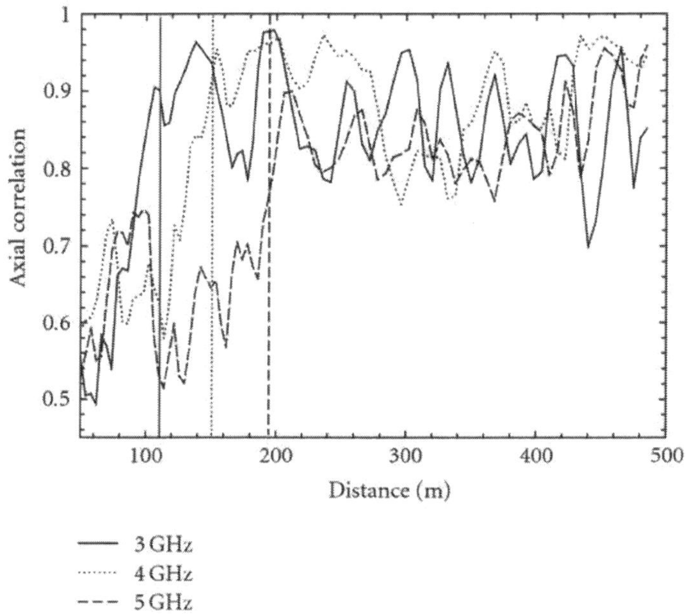

Figure 6. AXIAL CORRELATION BETWEEN RECEIVING ARRAYS 4 M (FOR 50 M < **d** < 202 M) OR 6 M (FOR 202 M < **d** < 500 M) APART FOR THREE FREQUENCIES AND THEIR CORRESPONDING BREAKPOINTS.

In Figure 6, the three vertical lines correspond to the positions of the breakpoint between the two slopes, for the three frequencies, respectively. This breakpoint thus occurs at distance $d_{breakpoint\_axial}$ from the transmitter and by plotting all curves for frequencies between 2.8 and 5 GHz, the following empirical formula giving has been obtained:

$$10\log_{10}(d_{breakpoint\_axial}) = 16.8 + 1.2f\ (GHz). \quad (4)$$

At the breakpoint and beyond this distance, the average correlation between the fields received by the array elements, 6 m apart, is equal to $\rho_{breakpoint} = 0.88$, with a standard deviation $\sigma_{\rho\_axial} = 0.06$, this result remaining valid in all the frequency range. In the first zone, that is, for $d < d_{breakpoint\_axial}$, the average variation of ρ_{axial} is modelled by

$$\langle \rho_{\text{axial}} \rangle = \rho_{\text{breakpoint}} + 0.06 \left(d_{\text{breakpoint\_axial}} - d \right) \tag{5}$$

the standard deviation $\sigma_{\rho\_\text{axial}}$ being also equal to 0.06.

This leads to the following expression for modeling the variation of the correlation coefficient along the tunnel axis:

$$\rho_{\text{axial,model}} = \langle \rho_{\text{axial}} \rangle + X_{\rho\_\text{axial}}. \tag{6}$$

4. FIELD DISTRIBUTION IN THE TRANSVERSE PLANE

4.1. Field Distribution Function

In the transverse plane, the field distribution was first studied by considering, for a given axial distance *d*, the 12×12 possible combinations of the Tx and Rx antennas, and 7 close frequencies, within a 10 MHz band, as earlier explained. This has been done for various frequency bands between 2.8 and 5 GHz. We have compared the measured data to those given by a Rayleigh, Weibull, Rician, Nakagami and Lognormal distribution, and then using the Kolmogorov-Smirnov [20] test to decide what distribution best fits the experimental results. A Rice distribution appears to be the optimum one, whatever the frequency. The mathematical expression of its probability density function (PDF) is given by

$$f(x \mid \nu, \sigma_{\text{RICE}}) = \frac{x}{\sigma_{\text{RICE}}^2} \exp\left(\frac{-(x^2+\nu^2)}{2\sigma_{\text{RICE}}^2} \right) I_0\left(\frac{x\nu}{\sigma_{\text{RICE}}^2} \right) \tag{7}$$

In this formula, $I_0(\cdot)$ is the modified Bessel function of the first kind with order zero and ν and σ_{RICE} are parameters to be adjusted. The first order moment is expressed as

$$E(x) = \sqrt{\frac{\pi}{2}} \sigma_{\text{RICE}} L_{1/2}\left(-\frac{\nu^2}{2\sigma_{\text{RICE}}^2} \right) = \sqrt{\frac{\pi}{2}} \sigma_{\text{RICE}} L_{1/2}(-K), \tag{8}$$

$L_{1/2}$ being a Laguerre polynomial.

Before explaining how the two parameters of the Rice distribution have been found, let us recall that, in the mobile communication area, a Rice distribution usually characterizes the field distribution in line of sight (LOS) conditions and in presence of a multipath propagation. Usually a *K* factor is introduced and defined as the ratio of signal power in dominant component, corresponding to the power of the direct ray, over the scattered, reflected power. One can follow the same approach by defining a *K* factor in a given receiving zone which is, in our case, defined by the segment 33 cm long in the transverse plane of the tunnel, along which measurements were carried out.

4.2. Ricean K Factor

Knowing the $(12, 12)$ matrix whose elements are the S_{21} complex values for successive positions of the Tx and Rx antennas in the transverse plane, one can calculate *K* at a distance *d* and a frequency *f*, from the following expression:

$$K = \frac{|\langle S_{21}\rangle|^2}{\langle |S_{21} - \langle S_{21}\rangle|^2\rangle}. \tag{9}$$

It must be clearly outlined that, in a tunnel, the *K* factor cannot be easily interpreted. Indeed, there is no contribution of random components to the received power, the position of the 4 reflecting walls being invariant. *K* could be related to richness in terms of propagation modes having a significant power in the receiving transverse plane, a high number of modes giving rise to a high fluctuating field. However, quantifying the relationship between *K* and mode richness is not easy since the field fluctuation depends not only on the amplitude of the modes but also on their relative phase velocity. In a tunnel, one can conclude that *K* just gives an indication on the relative range of variation of the received power in a given zone.

Curves in Figure 7 have been plotted for 2 frequencies: 3 and 5 GHz. In the transverse zone of investigation (33 cm), for distances smaller than 200 m, the *K* factor is below −15 dB, which means that the received power strongly varies in the transverse plan, nearly following a Rayleigh distribution. However, *K* increases with distance and reaches 0 dB or more beyond

400 m, the constant part of the distribution becoming equal to or greater than the random part.

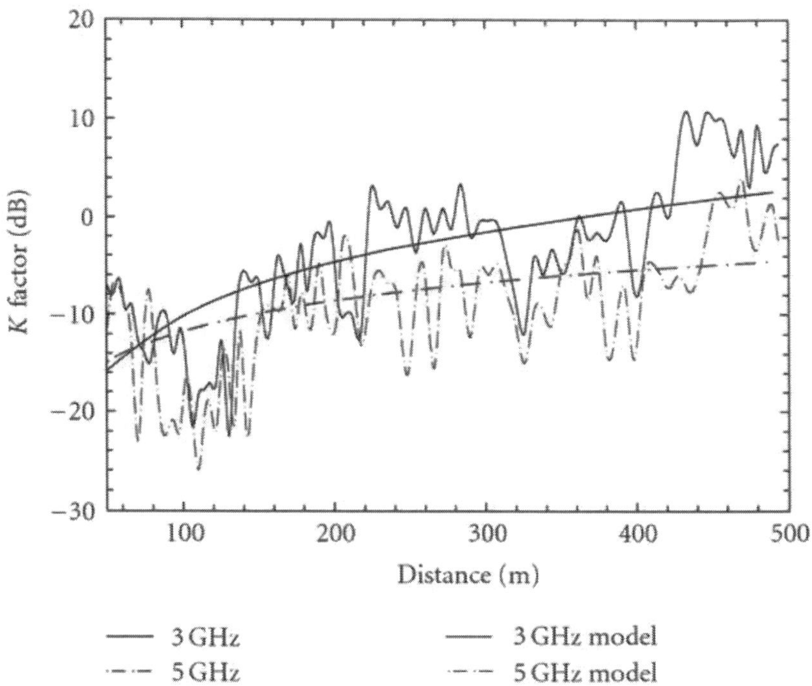

Figure 7. VARIATION OF THE **K** FACTOR AT 3 AND 5 GHZ, VERSUS DISTANCE, AND DEDUCED FROM MEASUREMENTS. Its average variation calculated from an empirical mathematical expression is also plotted.

This increase of *K* is due to the fact that the contribution of high-order modes becomes less important leading to less fluctuation of the transverse field. The same interpretation based on the modes can be made to interpret the influence of frequency on the *K* values. The variation of *K* is of course related to the variation of the correlation coefficient along the tunnel axis, as described in the previous section.

By following the same approach as for the path loss, described in Section 3, and thus by averaging *K* over groups of 7 frequencies and over 144 successive combinations of the transverse positions of the transverse positions of the Tx and Rx antennas, an empirical expression of the average *K* factor in terms of frequency and distance can be found. It is given by

$$\overline{K} = (K_0 + 10 \cdot n_{K_0} \log_{10}(f(\text{GHz})) + 10 \cdot (n_0 + 10 \cdot n_{n_0} \log_{10}(f(\text{GHz})))\log_{10}(d). \quad (10)$$

The best fit between the results given by (10) and those extracted from the measurements was obtained for the values of the parameters given in Table 2. The standard deviation between (10) and the measured *K* is given by σ_K.

Table 2. Parameters to be introduced in (10) for modeling the variation of the *K* factor.

	K_0	n_{K_0}	n_0	n_{n_0}	σ_K
Values	−79	6.73	3.6	−0.37	4.9

The curves labelled "model" in Figure 7 have been obtained by applying (10) and the above values for the parameters.

Let *X* be a random variable of zero mean. To completely describe the model, we can add to $\overline{K}$ such a random variable with a standard deviation of σ_K and labeled X_{σ_K}:

$$K_{\text{model}} = \overline{K} + X_{\sigma_K}. \quad (11)$$

4.3. Determination of the Ricean Parameters and Modeling of the Field Variation in the Transverse Plane

$\overline{K}$ is related to the field distribution parameters of the Rice distribution by

$$\overline{K} = \frac{\nu^2}{2\sigma^2_{\text{RICE}}}. \quad (12)$$

The mean value of $\overline{K}$ is deduced from (10) for a given frequency and distance, and by assuming a mean value of 1 of the amplitude of the field distribution $E(x) = 1$, the field distribution parameters ν and σ_{RICE} can be calculated. Note that mean value of the field would be determined by the large-scale fading, and fast variations around the mean value by the Rice distribution. Therefore, σ_{RICE} can be computed using (8) and (12):

$$\sigma_{\text{RICE}} = \sqrt{\frac{2}{\pi}} \frac{1}{L_{1/2}(-\overline{K})}. \tag{13}$$

By knowing σ_{RICE}, ν is immediately deduced from (12). As an example, curves (a) and (b) in Figure 8 compare the PDFs deduced from the measurements to those assuming a Rice distribution, for $d = 50$ m and $f = 5$ GHz, and $d = 500$ m and $f = 3$ GHz, respectively. We see the rather good agreement between measurements and the empirical formulation; the confidence level of the Smirnoff-Kolmogorov test remaining below 0.05.

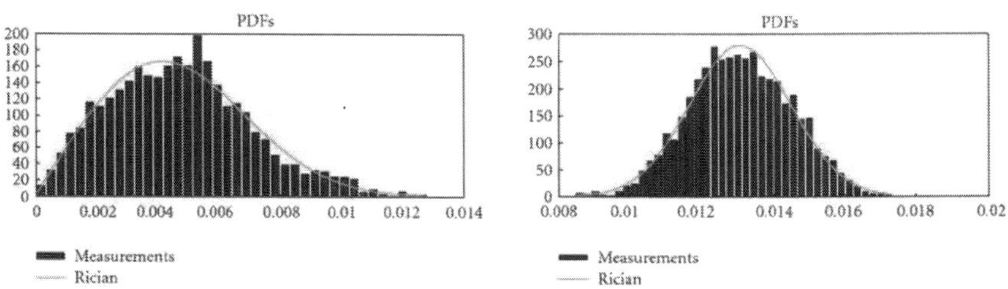

Figure 8 PDFS OF THE FIELD AMPLITUDE IN A TRANSVERSE PLANE EITHER DEDUCED FROM MEASUREMENTS OR CALCULATED ASSUMING A RICE DISTRIBUTION: (A) $d = 50$ m and $f = 5$ GHz, (b) $d = 500$ m and $f = 3$ GHz.

5. TRANSVERSE SPATIAL CORRELATION

The knowledge of the spatial correlation in the transverse plane is of special interest for MIMO systems. It is assumed, for simplicity, that the correlations at the transmitter and at the receiver are separable [21]. Furthermore, since the Rx and Tx antenna arrays are situated in the same transverse zone of the tunnel, one can expect that the correlation statistics are the same for the Tx site and for the Rx site and thus, in the following, they are not differentiated.

For each axial distance *d*, and for each frequency *f*, the amplitude of the complex correlation function ρ_{trans} was deduced from the 12×12 channel matrix, whose elements are associated to the successive positions of the Tx and Rx antennas in the transverse plane. Let *s* be the spacing between two receiving points. Figure 9 shows, for $f = 3$ GHz, the variation of ρ_{trans} versus the axial distanceand for different values of *s*: 3, 9, 21, and 33 cm.

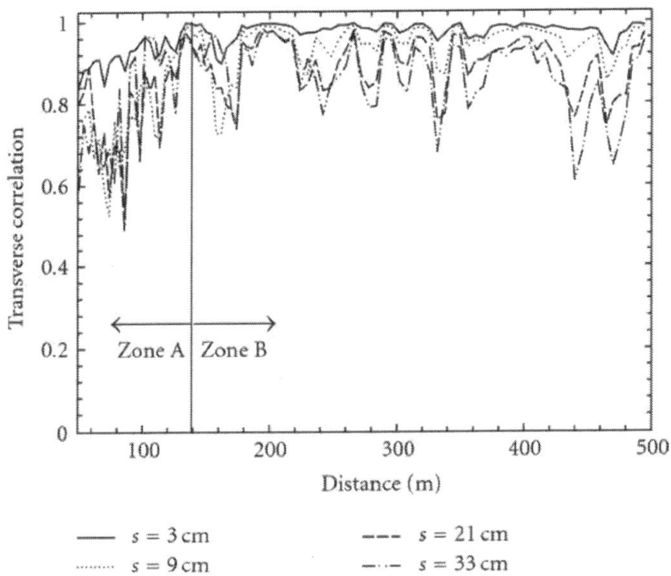

Figure 9. TRANSVERSE CORRELATION AT 3 GHZ VERSUS AXIAL DISTANCE AND FOR DIFFERENT SPACING IN THE TRANSVERSE PLANE.

ρ_{trans} is of course a decreasing function of the antenna spacing. Furthermore, for a given spacing, the correlation in the transverse plane increases when the axial distance increases, at least until the end of a zone, named *A* in Figure 9, occurring at a point called "breakpoint trans." This remark is connected to the comments made in Section 4 concerning the axial correlation, where we have outlined that, when the axial distance increases, the high-order modes are more and more attenuated, leading to a less fluctuating electromagnetic field. Beyond the "breakpoint trans" (zone *B* in Figure 9), ρ_{trans} keeps an average high value, even if local decreases are observed. The local decreases can be explained by the field pattern in the transverse plane of the tunnel. Indeed, this pattern does not present translation symmetry since it results from the combining of many modes, both in amplitude and in phase.

By analyzing results in the whole frequency range, it appears that the width of zone *A* slightly increases with frequency, as it occurred in the case of the longitudinal correlation (Section 4). Again, using all measured frequencies, an empirical formula giving the position of the "breakpoint_trans" point is given by

$$10 \log_{10}\left(d_{\text{breakpoint\_trans}}\right) = 16 + 1.7f\ (\text{GHz}). \tag{14}$$

In zone *B*, one can calculate the mean value $\rho_{\text{trans}}(s, \text{zone } B, f)$ by averaging $\rho_{\text{trans}}(s, d, f)$ over the axial distance *d*. The results are the curves plotted in Figure 10, versus frequency and for different values of *s*: 3, 9, 21, and 33 cm.

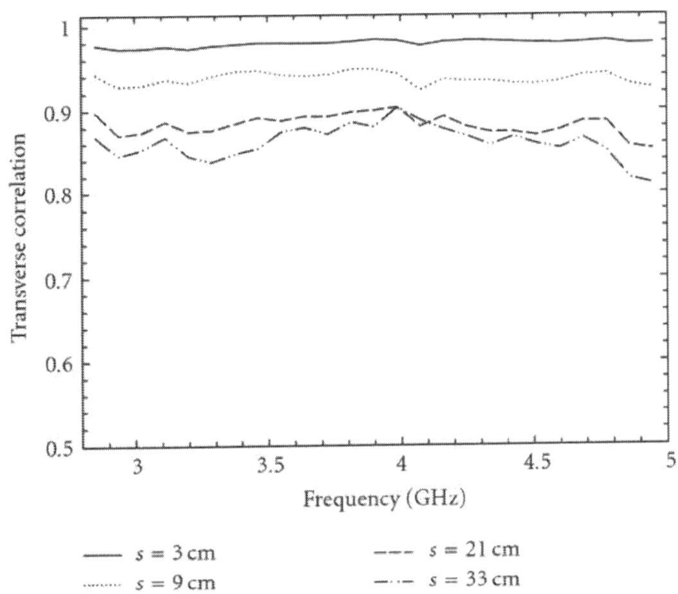

Figure 10. AVERAGE CORRELATION IN ZONE **B** , VERSUS FREQUENCY, AND FOR FOUR ANTENNA SPACING.

It appears that $\rho_{\text{trans}}(s, \text{zone } B, f)$ is nearly frequency independent and that an empirical formula fitting the experimental results can be obtained:

$$\rho_{\text{trans}}(s, \text{zone } B) = 0.98 - 0.0042s \text{ (cm)}. \tag{15}$$

The difference between (15) and the measured correlation is a random variable of zero mean and standard deviation $\sigma_{\rho\_\text{trans}}$:

$$\sigma_{\rho\_\text{trans}}(s, \text{zone } B) = 0.02 + 0.0025s \text{ (cm)}. \tag{16}$$

The modeling of the ρ_{trans} in zone *A* assumes an average linear variation with distance. The adequate formula in this zone is

$$\rho_{\text{trans}}(s, d, f) = \rho_{\text{trans}}(s, \text{zone } B) + 0.04\left(d_{\text{breakpoint\_trans}} - d\right). \tag{17}$$

In this formula, the implicit dependence on frequency comes from the value of $d_{\text{breakpoint\_trans}}$. The standard deviation around this value is nearly frequency independent and is modeled by

$$\sigma_{\rho_{\text{trans}}}(s) = 0.0324 + 0.0033s \text{ (cm)}. \tag{18}$$

Finally, for a given antenna spacing, the correlation between two antenna elements is modeled by

$$\rho_{\text{trans,model}} = \rho_{\text{trans}} + X_{\sigma\rho\_\text{trans}}. \tag{19}$$

6. FULL MODEL

The previous sections have proposed empirical formulas, based on experimental results, to model the path loss and the field fluctuation and correlation in a transverse plane. These formulas can be applied to randomly generate the transfer matrices **H** of an MIMO link in a straight tunnel having an arched cross-section, which is the shape of most road and railway tunnels. The transmitting and receiving arrays are supposed to be linear arrays, whose axes are horizontal and situated in the transverse plane of the tunnel, this configuration being quite usual. An approach based on the Kronecker model [21] was chosen for its simplicity.

To determine the various elements of **H**, the following steps can be followed:

1. define the system parameters, such as frequency, distance between the transmitter and the receiver, number of array elements at the transmitter and at the receiver, element spacing and number of snapshots, corresponding to the number of realizations to be simulated;

2. determine a value for the path loss $PL(f,d)$ using (3);

3. compute a K factor from (11). We recall that in (3) and in (11), the value given by the model is the sum of two terms: a deterministic one plus a random variable whose standard deviation is known;

4. knowing K and $PL(f,d)$, the elements of a $\mathbf{G}_{trans}$ matrix, having the same size as **H**, are randomly chosen in a normalized Ricean distribution;

5. as mentioned in Section 4, it was assumed that the correlations between either the transmitting elements or the receiving elements follow the same distribution. The terms of the correlation matrices at the transmitting and receiving sites, R_{Rx} and R_{Tx}, are thus deduced from (19).

The Kronecker model leads to

$$\mathbf{H} = \mathrm{PL}(f,d)\mathbf{R}_{\mathrm{Rx}}^{1/2}\mathbf{G}_{\mathrm{trans}}\left(\mathbf{R}_{\mathrm{Tx}}^{1/2}\right)^{T}. \tag{20}$$

To give an example of application of this formula, let us consider a 4×4 MIMO system at 4 GHz, an array element spacing of 0.8 λ (6 cm at 4 GHz) and a distance d between the transmitter and the receiver of 250 m.

The channel capacity of a MIMO system for a given channel realization **H** can be computed as [22]

$$C = \log_2\left(\det\left|\mathbf{I}_N + \frac{\mathrm{SNR}}{M}\mathbf{H}\mathbf{H}^{\dagger}\right|\right), \tag{21}$$

where $\mathbf{I}_N$ is the $N \times N$ identity matrix, $(\,)^{\dagger}$ is the transpose conjugate operation and SNR is the signal-to-noise ratio at the receiver. The channel capacity C was calculated by assuming a fixed SNR equal to 10 dB. A constant SNR was chosen because we want to emphasize the influence of correlation and field distribution in the transverse plane. To compute the capacity assuming a fixed transmitting power, the contribution of the path loss must be added, which is straightforward.

The model was applied by considering 1000 realizations and the cumulative probability density function of the capacity is plotted in Figure 11 (curve "model"). To be able to compare this distribution to experimental results, a large number of measured values are needed. To increase this number we have thus calculated the capacity not only at 4 GHz, but also for all frequencies within a 100 MHz band around 4 GHz. We see in Figure 11, the rather good agreement between results deduced from the experiments (curve "measurements") and those given by the model.

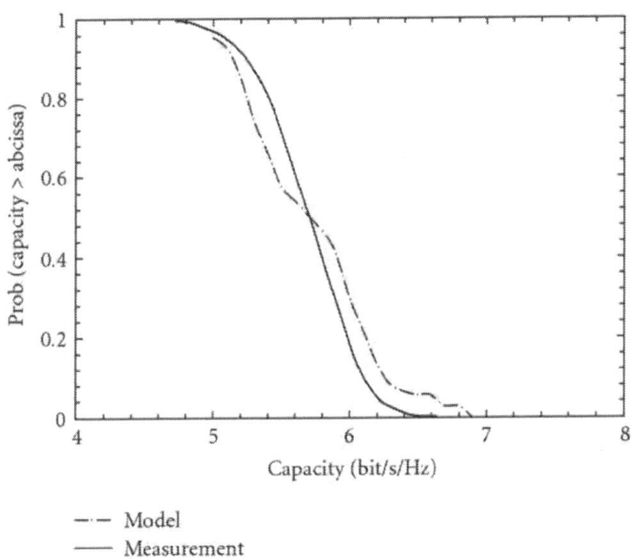

Figure 11. APPLICATION OF THE MIMO MODEL FOR A 4×4 MIMO system, for a frequency of 4 GHz and for a distance of 250 m.

7. CONCLUSION

The statistics of the electromagnetic field variation in a tunnel has been deduced from measurements made in an arched tunnel, which is the usual shape of road and railway tunnels, and in a frequency range extending from 2.8 to 5 GHz. Both the methodology of the experiments and the analysis were aimed at predicting the performance of an MIMO link in a wide frequency band.

It was shown, by subtracting the antenna effect, that the path loss is not strongly dependent on frequency and that the attenuation constant keeps small values, the tunnel behaving as a low-loss guiding structure. Along the investigated transverse axis of the tunnel, over 33 cm long, the small-scale fading follows a Ricean distribution. However, for distances between the transmitting and receiving antennas up to 200 m, the K factor is below −15 dB, meaning that the field is nearly Rayleigh distributed. It also appeared that K is an increasing function of distance, reaching 0 dB at about 400 m.

Empirical formulas to model the main propagation characteristics were proposed and applied to generate transfer matrices of an MIMO link.

ACKNOWLEDGMENTS

This work has been supported by the European FEDER funds, the Region Nord-Pas de Calais, and the French ministry of research, in the frame of the CISIT project.

REFERENCES

1. Yamaguchi Y, Abe T, Sekiguchi T: Radio wave propagation loss in the VHF to microwave region due to vehicles in tunnels. IEEE Transactions on Electromagnetic Compatibility 1989, 13(1):87-91.

2. Lienard M, Degauque P: Natural wave propagation in mine environments. IEEE Transactions on Antennas and Propagation 2000, 48(9):1326-1339.

3. Didascalou D, Maurer J, Wiesbeck W: Subway tunnel guided electromagnetic wave propagation at mobile communications

frequencies. IEEE Transactions on Antennas and Propagation 2001, 49(11):1590-1596.

4. Yang X, Lu Y: Research on propagation characteristics of millimeter wave in tunnels. International Journal of Infrared and Millimeter Waves 2007, 28(10):901-909.

5. Zhang YP, Hwang Y: Characterization of UHF radio propagation channels in tunnel environments for microcellular and personal communications. IEEE Transactions on Vehicular Technology 1998, 47(1):283-296.

6. Zhang YP, Zheng GX, Sheng JH: Radio propagation at 900 MHz in underground coal mines. IEEE Transactions on Antennas and Propagation 2001, 49(5):757-762.

7. Briso-Rodriguez C, Cruz JM, Alonso JI: Measurements and modeling of distributed antenna systems in railway tunnels. IEEE Transactions on Vehicular Technology 2007, 56(5, part 2):2870-2879.

8. Wang T-S, Yang C-F: Simulations and measurements of wave propagations in curved road tunnels for signals from GSM base stations. IEEE Transactions on Antennas and Propagation 2006, 54(9):2577-2584.

9. Ching GS, Ghoraishi M, Lertsirisopon N, et al.: Analysis of DSRC service over-reach inside an arched tunnel. IEEE Journal on Selected Areas in Communications 2007, 25(8):1517-1525.

10. Boutin M, Benzakour A, Despins CL, Affes S: Radio wave characterization and modeling in underground mine tunnels. IEEE Transactions on Antennas and Propagation 2008, 56(2):540-549.

11. Molina-Garcia-Pardo J-M, Lienard M, Degauque P, Dudley DG, Juan-Llàcer L: Interpretation of MIMO channel characteristics in rectangular tunnels from modal theory. IEEE Transactions on Vehicular Technology 2008, 57(3):1974-1979.

12. Liénard M, Degauque P, Baudet J, Degardin D: Investigation on MIMO channels in subway tunnels. IEEE Journal on Selected Areas in Communications 2003, 21(3):332-339.

13. Lienard M, Degauque P: Propagation in wide tunnels at 2 GHz: a statistical analysis. IEEE Transactions on Vehicular Technology 1998, 47(4):1322-1328.

14. Molisch AF: Ultrawideband propagation channels-theory, measurement, and modeling. IEEE Transactions on Vehicular Technology 2005, 54(5):1528-1545.

15. Rappaport TS: Wireless Communications. Prentice-Hall, Englewood Cliffs, NJ, USA; 1996.

16. Ghassemzadeh SS, Jana R, Rice CW, Turin W, Tarokh V: Measurement and modeling of an ultra-wide bandwidth indoor channel. IEEE Transactions on Communications 2004, 52(10):1786-1796.

17. Keignart J, Daniele N: Channel sounding and modeling for indoor UWB communications. Proceedings of International Workshop on Ultra Wideband Systems (IWUWBS '03), June 2003, Oulu, Finland

18. Cassioli D, Win MZ, Molisch AF: The ultra-wide bandwidth indoor channel: from statistical model to simulations. IEEE Journal on Selected Areas in Communications 2002, 20(6):1247-1257.

19. Hovinen V, Hämäläinen M, Tesi R, Hentilä L, Laine N: A proposal for a selection of indoor UWB path loss model. Wisair, Tel Aviv, Israel; 2002.http://grouper.ieee.org/groups/802/15/pub/2002/Jul02

20. Papoulis A, Pillai SU: Probability, Random Variables and Stochastic Processes. 4th edition. McGraw-Hill, Boston, Mass, USA; 2002.

21. Kermoal JP, Schumacher L, Pedersen KI, Mogensen PE, Frederiksen F: A stochastic MIMO radio channel model with experimental validation. IEEE Journal on Selected Areas in Communications 2002, 20(6):1211-1226.

22. Foschini GJ, Gans MJ: On limits of wireless communications in a fading environment when using multiple antennas. Wireless Personal Communications 1998, 6(3):311-335.

11 CHAPTER

PROPAGATION IN LOSSY RECTANGULAR WAVEGUIDES

Kim Ho Yeap[1], Choy Yoong Tham[1], Ghassan Yassin[1] and Kee Choon Yeong[1]

[1] *Tunku Abdul Rahman University, 1,2Malaysia*

1. INTRODUCTION

In millimeter and submillimeter radio astronomy, waveguide heterodyne receivers are often used in signal mixing. Wave guiding structures such as circular and rectangular waveguides are widely used in such receiver systems to direct and couple extraterrestrial signals at millimeter and submillimeter wavelengths to a mixer circuit (Carter et al., 2004; Boifot et al., 1990; Withington et al., 2003).

To illustrate in detail the applications of waveguides in receiver systems, a functional block diagram of a typical heterodyne receiver in radio telescopes is shown in Fig. 1 (Chattopadhyay et al., 2002). The electromagnetic signal (RF signal) from the antenna is directed down to the front end of the receiver system via mirrors and beam waveguides (Paine et al., 1994). At the front end of the receiver system, such as the sideband separating receiver designed for the ALMA band 7 cartridge (Vassilev and Belitsky, 2001a; Vassilev and Belitsky, 2001b; Vassilev et al., 2004), the RF signal is channelled from the aperture of the horn through a circular and subsequently a rectangular waveguide, before being coupled to the mixer. In the mixer circuit, a local oscillator (LO) signal which is generally of lower frequency is then mixed with the RF signal, to down convert the RF signal to a lower intermediate frequency (IF) signal. Here, a superconductor-insulator-superconductor (SIS) heterodyne mixer is commonly implemented for the process of down conversion. At the back end of the system, the IF signal goes through multiple stages of amplification and is, eventually, fed to a data analysis system such as an acousto-optic spectrometer. The data analysis system will then be able to perform Fourier transformation and record spectral information about the input signal.

The front-end receiver noise temperature T_R is determined by a number of factors. These include the mixer noise temperature T_M, the conversion loss C_{Loss}, the noise temperature of the first IF amplifier T_{IF}, and the coupling efficiency between the IF port of the junction and the input port of the first IF amplifier T_{IF}, . A comparison of the performance of different SIS waveguide receivers is listed in Table 1 (Walker et al., 1992). It can be seen that the value of T_R for the 230 GHz system is a factor of 3 to 4 less than that achieved with the 492 GHz system. The decrease in system performance at 492 GHz is due to the increase of C_{Loss} and T_M by a factor of approximately 3.

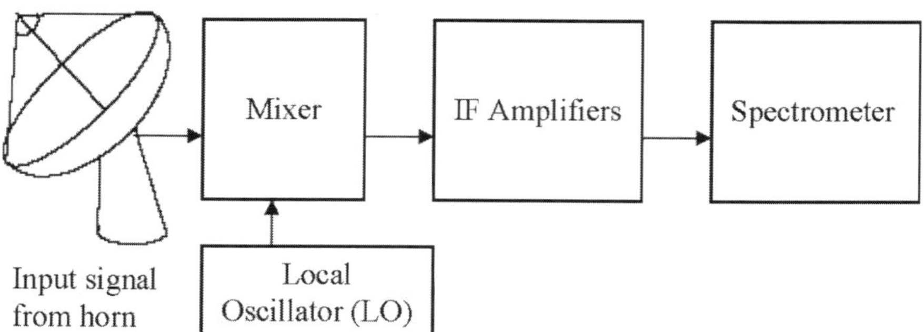

Figure 1. Block diagram of a heterodyne receiver.

Table 1. Comparison of SIS receiver performance.

SIS Junction	Nb	Pb	Nb
Center Frequency (GHz)	230	345	492
T_R (K)	48	159	176
T_M (K)	34	129	123
C_{Loss} (dB)	3.1	8.1	8.9
T_{IF} (K)	7.0	4.2	6.8

Since the input power level of the weak millimeter and submillimeter signals is quite small – i.e. of the order of 10^{-18} to 10^{-20} W (Shankar, 1986), it is therefore of primary importance to minimize the conversion loss C_{Loss} of the mixer circuit. One way of doing so, is to ensure that the energy of the LO and, in particular, the RF signals is channelled and coupled from the waveguides to the mixer circuit in a highly efficient manner. It is simply too time consuming and too expensive to develop wave guiding structures in a receiver system on a trial-and-error basis. To minimize the loss of the propagating signals, the availability of an accurate and easy-to-use mathematical model to compute the loss of such signals in wave guiding structures is, of course, central to the development of receiver circuits.

2. RELATED WORK

Analysis of the propagation of wave in circular cylindrical waveguides has already been widely performed (Glaser, 1969; Yassin et al., 2003; Claricoats, 1960a; Claricoats, 1960b; Chou and Lee, 1988). The analyses by these authors are all based on the rigorous method formulated by Stratton (1941). In Stratton's formulation, the fields at the wall surface are made continuous into the wall material. Assumption made on the field decaying inside the wall material yields relations which allow the propagation constant to be determined. Due to the difficulty in matching the boundary conditions in Cartesian coordinates, this approach, however fails to be implemented in the case of rectangular waveguides. A similar rigorous technique to study the attenuation of rectangular waveguides is not available hitherto.

The perturbation power-loss method has been commonly used in analyzing wave attenuation in lossy (Stratton, 1941; Seida, 2003; Collin 1991; Cheng, 1989) and superconducting (Winters and Rose, 1991; Ma, 1998; Wang et al., 1994; Yalamanchili et al., 1995) rectangular waveguides; respectively. This is partly due to its ability to produce simple analytical solution, and also partly because it gives reasonably accurate result at frequencies f well above its cutoff frequency f_c. In this method, the field expressions are derived by assuming that the walls to be of infinite conductivity. This allows the solution to be separated into pure Transverse Electric (TE) and Transverse Magnetic (TM) modes. For a waveguide with finite conductivity, however, a superposition of both TE and TM modes is necessary to satisfy the boundary conditions (Stratton, 1941; Yassin et al., 2003). This is because, unlike those of the lossless case, the modes in a lossy waveguide are no longer mutually orthogonal to each other (Collin, 1991). To calculate the attenuation, ohmic losses are assumed due to small field penetration into the conductor surface. Results however show that this method fails near cutoff, as the attenuation obtained diverges to infinity when the signal frequency f approaches the cutoff f_c. Clearly, it is more realistic to expect losses to be high but finite rather than diverging to infinity. The inaccuracy in the power-loss method at cutoff is due to the fact that the field equations are assumed to be identical to those of a lossless waveguide. Since a lossless waveguide behaves exactly like an ideal high pass filter, signals cease to propagate at f below f_c.

It can be seen that the assumption of lossless fields fail to give an insight or deeper understanding on the mechanism of the propagation of wave in practical lossy waveguides. Moreover, at very high frequency – especially that approaches the millimeter and submillimeter wavelengths – the loss tangent of the conducting wall decreases. Therefore, such assumption turns out to be inaccurate at very high frequency. Although Stratton (1941) has developed a truly fundamental approach to analyze waveguides, his approach is only restricted to the case of circular waveguides and could not be applied to rectangular waveguides. The workhorse of this chapter is, therefore, to develop a novel and accurate formulation – i.e. one that does not assume lossless boundary conditions – to investigate the loss of waves in rectangular waveguides. In particular, the new method shall be found more accurate and useful for waveguides operating at very high frequencies, such as those in the millimeter and submillimeter wavelengths. In Yeap et al. (2009), a simple method to compute the loss of waves in rectangular waveguides has been developed. However, the drawback of the method is that different sets of characteristic equation are required to solve for the propagation constants of different modes. Here, the method proposed in Yeap et al. (2009) shall be developed further so that the loss of different modes can be conveniently computed using only a single set of equation. For convenience purpose, the new method shall be referred to as the boundary-matching method in the subsequent sections.

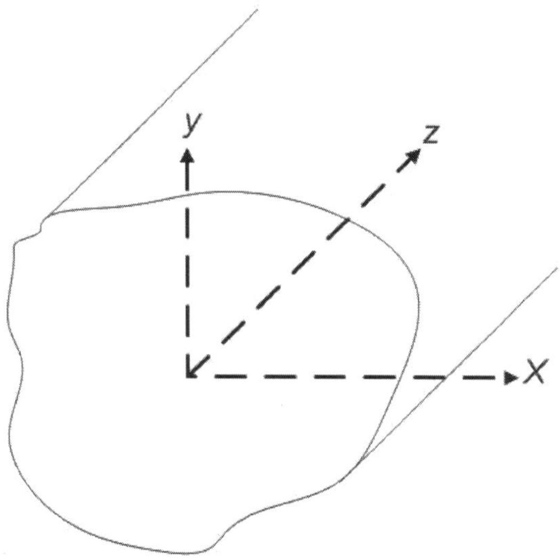

Figure 2. A waveguide with arbitrary geometry.

3. GENERAL WAVE BEHAVIOURS ALONG UNIFORM GUIDING STRUCTURES

As depicted in Fig. 2, a time harmonic field propagating in the z direction of a uniform guiding structure with arbitrary geometry can be expressed as a combination of elementary waves having a general functional form (Cheng, 1989)

$$\psi = \psi^0(x, y)\exp[j(\omega t + k_z z)] \tag{1}$$

where $\psi^0(x, y)$ is a two dimensional vector phasor that depends only on the cross-sectional coordinates, $\omega = 2\pi f$ the angular frequency, and k_z is the propagation constant.

Hence, in using phasor representation in equations relating field quantities, the partial derivatives with respect to t and z may be replaced by products with $j\omega$ and jk_z, respectively; the common factor $\exp[j(\omega t + k_z z)]$ can be dropped. Here, the propagation constant k_z is a complex variable, which consists of a phase constant β_z and an attenuation constant α_z

$$k_z = \beta_z - j\alpha_z \tag{2}$$

The field intensities in a charge-free dielectric region (such as free-space), satisfy the following homogeneous vector Helmholtz's equation

$$\nabla^2 \psi_z + (k^2 - k_z^2)\psi_z = 0 \tag{3}$$

where ψ_z is the longitudinal component of ψ, ∇^2 is the Laplacian operator for the transverse coordinates, and k is the wavenumber in the material. For waves propagating in a hollow waveguide, $k = k_0$, the wavenumber in free-space.

It is convenient to classify propagating waves into three types, in correspond to the existence of the longitudinal electric field E_z or longitudinal magnetic H_z field:

1. Transverse electromagnetic (TEM) waves. A TEM wave consists of neither electric fields nor magnetic fields in the longitudinal direction.

2. Transverse magnetic (TM) waves. A TM wave consists of a nonzero electric field but zero magnetic field in the longitudinal direction.

3. Transverse electric (TE) waves. A TE wave consists of a zero electric field but nonzero magnetic field in the longitudinal direction.

Single-conductor waveguides, such as a hollow (or dielectric-filled) circular and rectangular waveguide, cannot support TEM waves. This is because according to Ampere's circuital law, the line integral of a magnetic field around any closed loop in a transverse plane must equal the sum of the longitudinal conduction and displacement currents through the loop. However, since a single-conductor waveguide does not have an inner conductor and that the longitudinal electric field is zero, there are no longitudinal conduction and displacement current. Hence, transverse magnetic field of a TEM mode cannot propagate in the waveguide (Cheng, 1989).

4. FIELDS IN CARTESIAN COORDINATES

For waves propagating in a rectangular waveguide, such as that shown in Fig. 3, Helmholtz's equation in (3) can be expanded in Cartesian coordinates to give

$$\frac{\partial^2 \psi_z}{\partial x^2} + \frac{\partial^2 \psi_z}{\partial y^2} + \left(k^2 - k_z^{\,2}\right)\psi_z = 0 \tag{4}$$

By applying the method of separation of variables, ψ_z can be expressed as

$$\psi_z = X(x)Y(y) \tag{5}$$

Equation (4) can thus be separated into two sets of second order differential equations, as shown below (Cheng, 1989)

$$\frac{d^2 X(x)}{dx^2} + k_x^{\,2} X(x) = 0 \tag{6}$$

$$\frac{d^2 Y(y)}{dy^2} + k_y^{\,2} Y(y) = 0 \tag{7}$$

where k_x and k_y are the transverse wavenumbers in the *x* and *y* directions, respectively. The longitudinal fields can be obtained by solving (6) and (7) based on a set of boundary conditions and substituting the solutions into (5).

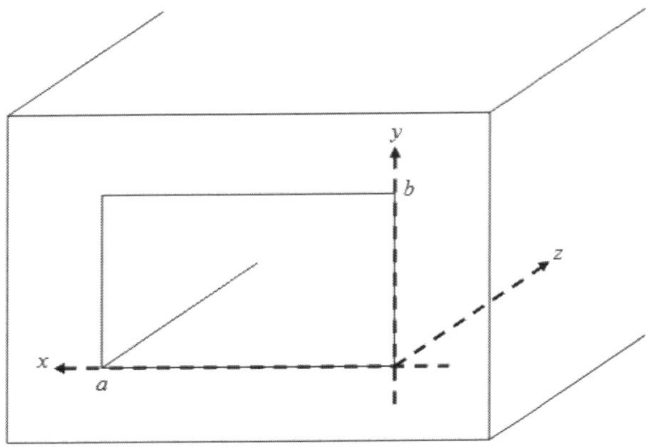

Figure 3. The cross section of a rectangular waveguide.

The transverse field components can be derived by substituting the longitudinal field components into Maxwell's source free curl equations

$$\nabla \times E = -j\omega\mu H \tag{8}$$

$$\nabla \times H = j\omega\varepsilon E \tag{9}$$

where ε and μ are the permittivity and permeability of the material, respectively. Expressing the transverse field components in term of the longitudinal field components E_z and H_z, the following equations can be obtained (Cheng, 1989)

$$H_x = \frac{-j}{k_x^{\,2} + k_y^{\,2}}\left(k_z \frac{dH_z}{dx} - \omega\varepsilon \frac{dE_z}{dy}\right) \tag{10}$$

$$H_y = \frac{-j}{k_x{}^2 + k_y{}^2}\left(k_z \frac{dH_z}{dy} + \omega\varepsilon \frac{dE_z}{dx}\right) \tag{11}$$

$$E_x = \frac{-j}{k_x{}^2 + k_y{}^2}\left(k_z \frac{dE_z}{dx} + \omega\mu \frac{dH_z}{dy}\right) \tag{12}$$

$$E_y = \frac{-j}{k_x{}^2 + k_y{}^2}\left(k_z \frac{dE_z}{dy} - \omega\mu \frac{dH_z}{dx}\right) \tag{13}$$

5. REVIEW OF THE POWER-LOSS METHOD

In the subsequent sections, analysis and comparison between the perturbation power-loss method and the new boundary matching method shall be performed. Hence, in order to present a complete scheme, the derivation of the conventional power-loss method is briefly outlined in this section.

The attenuation of electromagnetic waves in waveguides can be caused by two factors – i.e. the attenuation due to the lossy dielectric material $\alpha_{z(d)}$, and that due to the ohmic losses in imperfectly conducting walls $\alpha_{z(c)}$ (Cheng, 1989)

$$\alpha_z = \alpha_{z(d)} + \alpha_{z(c)} \,, \tag{14}$$

For a conducting waveguide, the inner core is usually filled with low-loss dielectric material, such as air. Hence, $\alpha_{z(d)}$ in (14) shall be assumed zero in the power-loss method and the loss in a waveguide is assumed to be caused solely by the conduction loss. It could be seen later that such assumption is not necessary in the new boundary-matching method. Indeed, the new method inherently accounts for both kinds of losses in its formulation.

The approximate power-loss method assumes that the fields' expression in a highly but imperfectly conducting waveguide, to be the same as those of a lossless waveguide. Hence, k_x, k_y, and k_z are given as (Cheng, 1989)

$$k_x = \frac{m\pi}{a} \tag{15}$$

$$k_y = \frac{n\pi}{b} \tag{16}$$

$$k_z = \beta_z \tag{17}$$

where a and b are the width and height, respectively, of the rectangular waveguide; whereas m and n denote the number of half cycle variations in the x and y directions, respectively. Every combination of m and n defines a possible mode for TE_{mn} and TM_{mn} waves.

Conduction loss is assumed to occur due to small fields' penetration into the conductor surfaces. According to the law of conservation of energy, the attenuation constant due to conduction loss can be derived as (Cheng, 1989)

$$\alpha_z = \frac{P_L}{2P_z} \tag{18}$$

where P_z is the time-average power flowing through the cross-section and P_L the time-average power lost per unit length of the waveguide.

Solving for P_L and P_z based on Poynting's theorem, the attenuation constant α_z for TM and TE modes – i.e. $\alpha_{z(TM)}$ and $\alpha_{z(TE)}$, respectively, can thus be expressed as (Collin, 1991)

$$\alpha_{z(TM)} = \frac{2R_s(m^2b^3 + n^2a^3)}{\eta ab\sqrt{1-\left(\frac{f_c}{f}\right)^2}\left(m^2b^2 + n^2a^2\right)} \tag{19}$$

$$\alpha_{z(TE)} = \frac{2R_s}{\eta b\sqrt{1-\left(\frac{f_c}{f}\right)^2}}\left\{\left(1+\frac{b}{a}\right)\left(\frac{f_c}{f}\right)^2 + \frac{b}{a}\left[1-\left(\frac{f_c}{f}\right)^2\right]\left[\frac{m^2ab + n^2a^2}{(mb)^2 + (na)^2}\right]\right\} \tag{20}$$

where R_s is the surface resistance, f_c the cutoff frequency, and η the intrinsic impedance of free space.

6. THE NEW BOUNDARY-MATCHING METHOD

It is apparent that, in order to derive the approximate power-loss equations illustrated in section 5, the field equations must be assumed to be lossless. In a lossless waveguide, the boundary condition requires that the resultant tangential electric field E_t and the normal derivative of the tangential magnetic field $\partial H_t/\partial a_n$ to vanish at the waveguide wall, where a_n is the normal direction to the waveguide wall. In reality, however, this is not exactly the case. The conductivity of a practical waveguide is finite. Hence, both E_t and $\partial H_t/\partial a_n$ are not exactly zero at the boundary of the waveguide. Besides, the loss tangent of a material decreases in direct proportion with the increase of frequency. Hence, a highly conducting wall at low frequency may exhibit the properties of a lossy dielectric at high frequency, resulting in inaccuracy using the assumption at millimeter and submillimeter wavelengths.

In order to model the field expressions closer to those in a lossy waveguide and to account for the presence of fields inside the walls, two phase parameters have been introduced in the new method. The phase parameters – i.e $_x$ and $_y$, are referred to as the field's penetration factors in the x and y directions, respectively. It is worthwhile noting that, with the introduction of the penetration factors, E_t and $\partial H_t/\partial a_n$ do not necessarily decay to zero at the boundary, therefore allowing the effect of not being a perfect conductor at the waveguide wall.

6.1. Fields in a lossy rectangular waveguide

For waves propagating in a lossy hollow rectangular waveguide, as shown in Figure 3, a superposition of TM and TE waves is necessary to satisfy the boundary condition at the wall (Stratton, 1941; Yassin et al., 2003). The longitudinal electric and magnetic field components E_z and H_z, respectively, can be derived by solving Helmholtz's homogeneous equation in Cartesian coordinate. Using the method of separation of variables (Cheng, 1989), the following set of field equations is obtained

$$E_z = E_0 \sin(k_x x + \phi_x)\sin(k_y y + \phi_y) \tag{21}$$

$$H_z = H_0 \cos(k_x x + \phi_x)\cos(k_y y + \phi_y) \tag{22}$$

where E_0 and H_0 are constant amplitudes of the fields.

The propagation constant k_z for each mode will be found by solving for k_x and k_y and substituting the results into the dispersion relation

$$k_z = \sqrt{k_0^2 - k_x^2 - k_y^2} \tag{23}$$

Equations (21)and (22) must also apply to a perfectly conducting waveguide. In that case E_z and $\partial H_z/\partial a_n$ are either at their maximum magnitude or zero at both x = a/2 and y = b/2 – i.e. the centre of the waveguide, therefore

$$\sin\left(\frac{k_x a}{2}+\phi_x\right)=\sin\left(\frac{k_y b}{2}+\phi_y\right)=\pm 1 \text{ or } 0 \tag{24}$$

Solving (24), the penetration factors are obtained as,

$$\phi_x=\frac{(m\pi-k_x a)}{2} \tag{25}$$

$$\phi_y=\frac{(n\pi-k_y b)}{2} \tag{26}$$

For waveguides with perfectly conducting wall, $k_x = m\pi/a$ and $k_y = n\pi/b$, (25) and (26) result in zero penetration and E_z and H_z in (21) and (22) are reduced to the fields of a lossless waveguide. To take the finite conductivity into account, k_x and k_y are allowed to take complex values yielding non-zero penetration of the fields into the waveguide material

$$k_x=\beta_x-j\alpha_x \tag{27}$$

$$k_y=\beta_y-j\alpha_y \tag{28}$$

where β_x and β_y are the phase constants and α_x and α_y are the attenuation constants in the x and y directions, respectively. Substituting the transverse wavenumbers in (27) and (28) into (23), the propagation constant of the waveguide k_z results in a complex value, therefore, yielding loss in wave propagation.

Substituting (21) and (22) into (10) to (13), the fields are obtained as

$$H_x=\frac{j\left[k_z k_x H_0+\omega\varepsilon_0 k_y E_0\right]\sin(k_x x+\varphi_x)\cos(k_y y+\phi_y)}{k_x{}^2+k_y{}^2} \tag{29}$$

$$H_y = \frac{j\left[k_z k_y H_0 - \omega\varepsilon_0 k_x E_0\right]\cos(k_x x + \phi_x)\sin(k_y y + \phi_y)}{k_x{}^2 + k_y{}^2}$$

(30)

$$E_x = -\frac{j\left[k_z k_x E_0 - \omega\mu_0 k_y H_0\right]\cos(k_x x + \phi_x)\sin(k_y y + \phi_y)}{k_x{}^2 + k_y{}^2}$$

(31)

$$E_y = -\frac{j\left[k_z k_y E_0 + \omega\mu_0 k_x H_0\right]\sin(k_x x + \phi_x)\cos(k_y y + \phi_y)}{k_x{}^2 + k_y{}^2}$$

(32)

where μ_0 and ε_0 are the permeability and permittivity of free space, respectively.

6.2. Formulation

At the wall, the tangential fields must satisfy the relationship defined by the constitutive properties μ_c and ε_c of the material. The ratio of the tangential component of the electric field to the surface current density at the conductor surface is represented by (Yeap et al., 2009b; Yeap et al., 2010)

where μ_c and ε_c are the permeability and permittivity of the wall material, respectively, and

$\sqrt{\dfrac{\mu_c}{\varepsilon_c}}$ is the intrinsic impedance of the wall material. The dielectric constant is complex and ε_c may be written as

$$\varepsilon_c = \varepsilon_0 - j\frac{\sigma_c}{\omega}$$

(34)

where σ_c is the conductivity of the wall.

In order to estimate the loss of waves in millimeter and submillimeter wavelengths more accurately, a more evolved model than the conventional constant conductivity model used at microwave frequencies is necessary. Here, Drude's model is applied for the frequency dependent conductivity σ_c (Booker, 1982)

$$\sigma_c = \frac{\sigma}{(1 + j\omega\tau)}$$

(35)

where σ is the conventional constant conductivity of the wall material and τ the mean free time. For most conductors, such as Copper, the mean free time τ is in the range of 10^{-13} to 10^{-14} s (Kittel, 1986).

At the width surface of the waveguide, $y = b$, $E_z/H_x = -E_x/H_z = \sqrt{\frac{\mu_c}{\varepsilon_c}}$.
Substituting (21), (22), (29), and (31) into (33), the following relationships are obtained

$$\frac{-E_x}{H_z} = \frac{j}{k_x^2 + k_y^2}\left(\frac{E_0}{H_0}k_z k_x - \omega\mu_0 k_y\right)\tan\left(k_y b + \phi_y\right) = \sqrt{\frac{\mu_c}{\varepsilon_c}}$$

$$\frac{H_x}{E_z} = \frac{j}{k_x^2 + k_y^2}\left(\frac{H_0}{E_0}k_z k_x + \omega\varepsilon_0 k_y\right)\cot\left(k_y b + \phi_y\right) = \sqrt{\frac{\varepsilon_c}{\mu_c}}$$

(36 a,b)

Similarly, at the height surface where x = a, we obtain $E_y/H_z = -E_z/H_y = \sqrt{\frac{\mu_c}{\varepsilon_c}}$ Substituting (21), (22), (30), and (32) into (33), the following relationships are obtained

$$\frac{E_y}{H_z} = \frac{-j}{k_x^2 + k_y^2}\left(\frac{E_0}{H_0}k_z k_y + \omega\mu_0 k_x\right)\tan(k_x a + \phi_x) = \sqrt{\frac{\mu_c}{\varepsilon_c}}$$

$$\frac{-H_y}{E_z} = \frac{-j}{k_x^2 + k_y^2}\left(\frac{H_0}{E_0}k_z k_y - \omega\varepsilon_0 k_x\right)\cot(k_x a + \phi_x) = \sqrt{\frac{\varepsilon_c}{\mu_c}}$$

(37 a,b)

In order to obtain nontrivial solutions for (36) and (37), the determinant of the equations must be zero (Yeap et al., 2009a). By letting the determinant of the coefficients of E_0 and H_0 in (36) and (37) vanish the following transcendental equations are obtained

$$\left[\frac{j\omega\mu_0 k_y \tan(k_y b + \phi_y)}{k_x^2 + k_y^2} + \sqrt{\frac{\mu_c}{\varepsilon_c}}\right]\left[\frac{j\omega\varepsilon_0 k_y \cot(k_y b + \phi_y)}{k_x^2 + k_y^2} - \sqrt{\frac{\varepsilon_c}{\mu_c}}\right] = \left[\frac{k_z k_x}{k_x^2 + k_y^2}\right]^2$$

$$\left[\frac{j\omega\mu_0 k_x \tan(k_x a + \phi_x)}{k_x^2 + k_y^2} + \sqrt{\frac{\mu_c}{\varepsilon_c}}\right]\left[\frac{j\omega\varepsilon_0 k_x \cot(k_x a + \phi_x)}{k_x^2 + k_y^2} - \sqrt{\frac{\varepsilon_c}{\mu_c}}\right] = \left[\frac{k_z k_y}{k_x^2 + k_y^2}\right]^2$$

(38 a,b)

Since the dominant TE_{10} mode has the lowest cutoff frequency among all modes and is the only possible mode propagating alone, it is of engineering importance. In the subsequent sections, comparison and detail analysis of the TE_{10} mode shall be performed. For the TE_{10} mode, m and n are set to 1 and 0, respectively. Substituting m = 1 and n = 0 into the penetration factors in (25) and (26), the transcendental equations in (38) for TE_{10} mode can be simplified to

$$\left[\sqrt{\frac{\mu_c}{\varepsilon_c}} + \frac{j\omega\mu_0 k_y \tan\left(k_y \frac{b}{2}\right)}{k_x^2 + k_y^2}\right]\left[\sqrt{\frac{\varepsilon_c}{\mu_c}} - \frac{j\omega\varepsilon_0 k_y \cot\left(k_y \frac{b}{2}\right)}{k_x^2 + k_y^2}\right] = \left[\frac{jk_z k_x}{k_x^2 + k_y^2}\right]^2$$

$$\left[\sqrt{\frac{\mu_c}{\varepsilon_c}} - \frac{j\omega\mu_0 k_x \cot\left(k_x \frac{a}{2}\right)}{k_x^2 + k_y^2}\right]\left[\sqrt{\frac{\varepsilon_c}{\mu_c}} + \frac{j\omega\varepsilon_0 k_x \tan\left(k_x \frac{a}{2}\right)}{k_x^2 + k_y^2}\right] = \left[\frac{jk_z k_y}{k_x^2 + k_y^2}\right]^2$$

(39 a,b)

In the above equations, k_x and k_y are the unknowns and k_z can then be obtained from the dispersion relation in (23). A multi root searching algorithm, such as the Powell Hybrid root searching algorithm in a NAG routine, can be used to find the roots of k_x and k_y. The routine requires initial guesses of k_x and k_y for the search. For good conductors, suitable guess values are clearly those close to the perfect conductor values i.e. $\beta_x = m\pi/a$, $\beta_y = n\pi/b$, $\alpha_x = \alpha_y = 0$. Hence, for the TE_{10} mode, the initial guesses for k_x and k_y are π/a and 0 respectively.)

6.3. Experimental setup

To validate the results, experimental measurements had been carried out. The loss as a function of frequency for a rectangular waveguide was measured using a Vector Network Analyzer (VNA). A 20 cm copper rectangular waveguide with dimensions of a = 1.30 cm and b = 0.64 cm such as that shown in Fig. 4 were used in the measurement.

Figure 4. Rectangular waveguides with width a = 1.30 cm and height b = 0.64 cm.

To minimize noise in the waveguide, a pair of chokes had also been designed and fabricated as shown in Fig. 5. A detail design of the choke drawn using AutoCAD are shown in Fig. 6(a) and Fig. 6(b).

In order to allow the waveguide to be connected to the adapters which are of different sizes, a pair of taper transitions had also been used as shown in Fig. 7. Fig. 8 depicts the complete setup of the experiment where the rectangular waveguide was connected to the VNA via tapers, chokes, coaxial cables, and adapters. Before measurement was carried out, the coaxial cables and waveguide adapters were calibrated to eliminate noise from the two devices. The loss in the waveguide was then observed from the S_{21} or S_{12} parameter of the scattering matrix. The measurement was performed in the frequencies at the vicinity of cutoff.

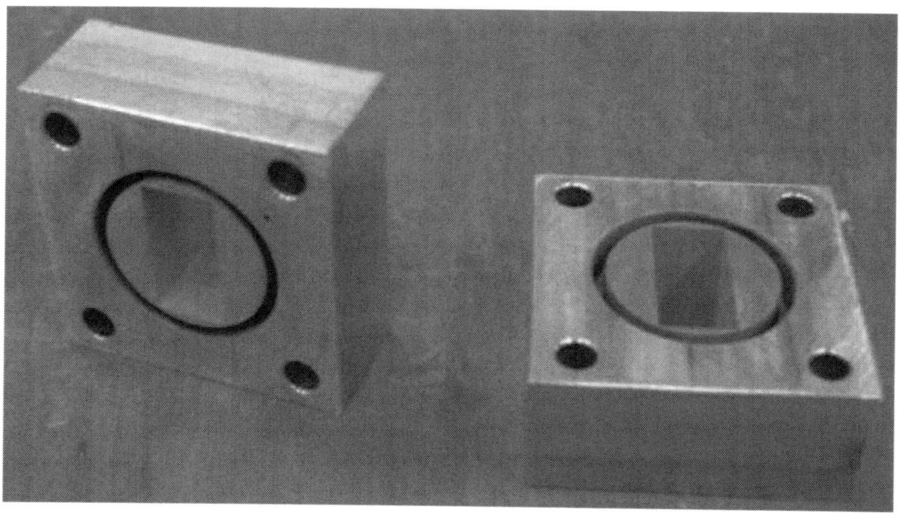

Figure 5. A pair of chokes made of aluminum.

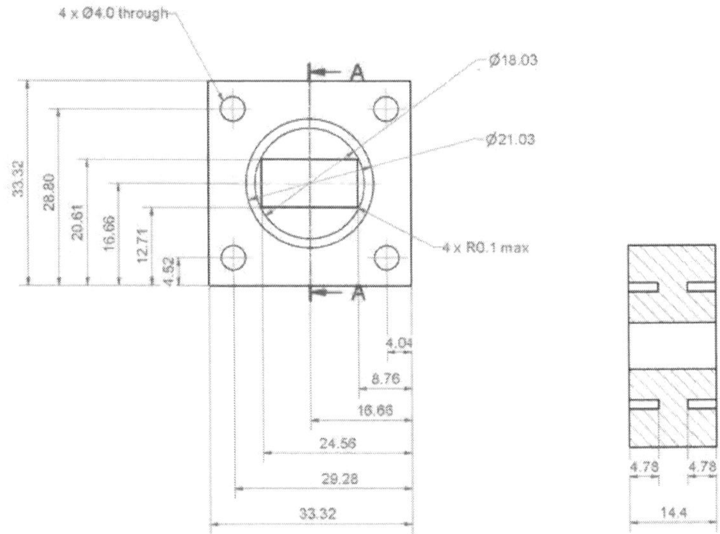

(a) (b)

Figure 6. Parameters of the (a) cross section and (b) side view of the choke.

Figure 7. Taper transitions.

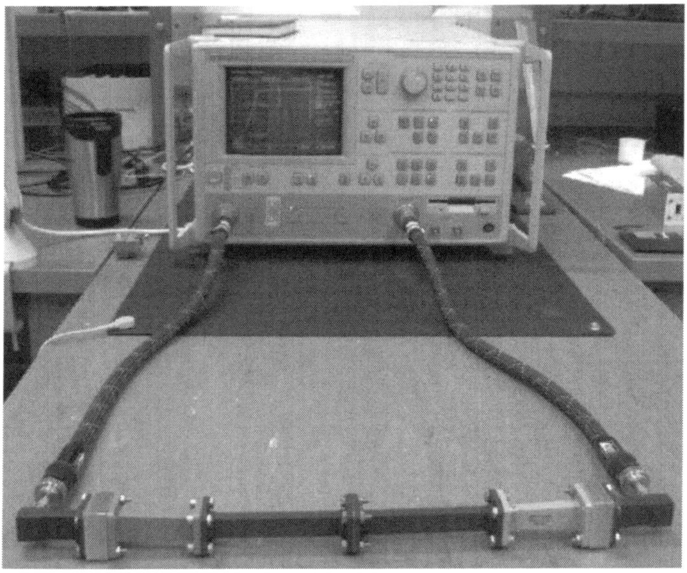

Figure 8. A 20 cm rectangular waveguide connected to the VNA, via tapers, chokes, adapters, and coaxial cables.

6.4. Results and discussion

As shown in Fig. 9, a comparison among the attenuation of the TE_{10} mode near cutoff as computed by the new method, the conventional power-loss method, and the measured S_{21} result was performed. Clearly, the attenuation constant α_z computed from the power-loss method diverges sharply to infinity, as the frequency approaches f_c, and is very different to the simulated results, which show clearly that the loss at frequencies below f_c is high but finite. The attenuation computed using the new boundary-matching method, on the other hand, matches very closely with the S_{21} curve, measured using from the VNA. As shown in Table 2, the loss between 11.47025 GHz and 11.49950 GHz computed by the boundary-matching method agrees with measurement to within 5% which is comparable to the error in the measurement. The inaccuracy in the power-loss method is due to the fact that the fields expressions are assumed to be lossless – i.e. k_x and k_y are taken as real variables. Analyzing the dispersion relation in (23), it could be seen that, in order to obtain α_z, k_x and/or k_y must be complex, given that the wavenumber in free space is purely real. Although the initial guesses for k_x and k_y applied in the new boundary-matching method are assumed to be identical with the lossless case, the final results actually converge to complex values when the characteristic equations are solved numerically.

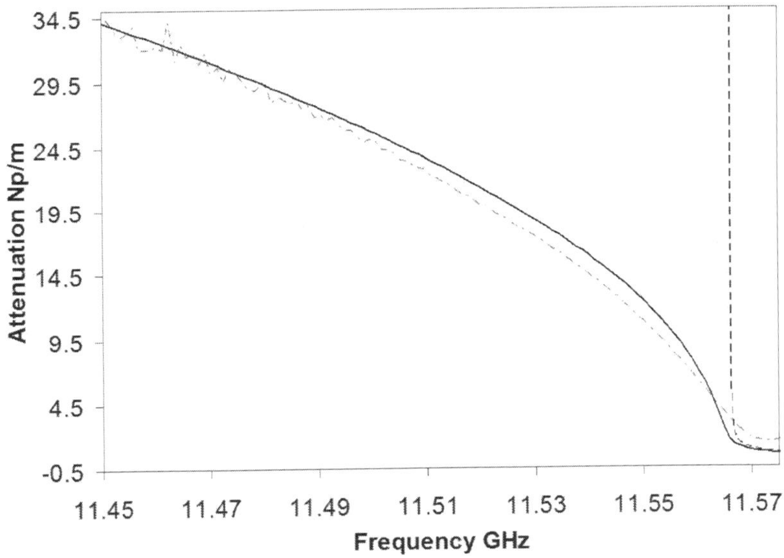

Figure 9. Attenuation of TE_{10} mode at the vicinity of cutoff. the new boundary matching method. power loss method. S21 measurement.

Fig. 10 shows the attenuation curve when the frequency is extended to higher values. Here, the loss due to TE_{10} alone could no longer be measured alone, since higher-order modes, such as TE_{11} and TM_{11}, etc., start to propagate. Close inspection shows that the loss predicted by the two methods at higher frequencies is in very close agreement. It is, therefore, sufficed to say that, although the power-loss method fails to predict the attenuation near f_c accurately, it is still considered adequate in computing the attenuation of TE_{10} in lossy waveguides, provided that the frequency f is reasonably above the cutoff f_c.

As depicted in Fig. 11, at frequencies beyond millimeter wavelengths, however, the loss computed by the boundary-matching method appears to be much higher than those by the power-loss method. The differences can be attributed to the fact that at extremely high frequencies, the loss tangent of the wall material decreases and the field in a lossy waveguide can no longer be approximated to those derived from a perfectly conducting waveguide. At such high frequencies, the wave propagating in the waveguide is a hybrid mode and the presence of the longitudinal electric field E_z can no longer be neglected.

Table 2. Attenuation of TE_{10} at the vicinity of the cutoff frequency.

Frequency GHz	Experiment	Boundary-matching method	%Δ
11.47025	30.17693	30.95782	2.59
11.47138	30.68101	30.77417	0.30
11.47250	29.53345	30.5894	3.58
11.47363	30.51672	30.40349	0.37
11.47475	30.16449	30.21642	0.17
11.47588	29.68032	30.02816	1.17
11.47700	29.09721	29.8387	2.55
11.47813	28.85077	29.648	2.76
11.47925	29.25528	29.45606	0.69
11.48038	29.20923	29.26283	0.18
11.48150	27.99881	29.06831	3.82
11.48263	28.38341	28.87245	1.72
11.48375	28.18551	28.67524	1.74
11.48488	27.91169	28.47664	2.02
11.48600	28.08407	28.27663	0.69
11.48713	27.44495	28.07517	2.30
11.48825	27.67956	27.87224	0.70
11.48938	26.84192	27.66779	3.08
11.49050	26.95767	27.46181	1.87
11.49163	26.60108	27.25425	2.46

11.48825	27.67956	27.87224	0.70
11.48938	26.84192	27.66779	3.08
11.49050	26.95767	27.46181	1.87
11.49163	26.60108	27.25425	2.46
11.49275	26.78715	27.04508	0.96
11.49388	26.14928	26.83426	2.62
11.49500	25.83003	26.62174	3.07
11.49613	25.82691	26.4075	2.25
11.49725	25.26994	26.19148	3.65
11.49838	24.82685	25.97365	4.62
11.49950	25.1100	25.75395	2.56

Unlike the power-loss method which only gives the value of the attenuation constant, one other advantage of the boundary-matching method is that it is able to account for the phase constant of the wave as well. A comparison between the attenuation constant and phase constant of a TE_{10} mode is shown in Fig. 12. As can be observed, as the attenuation in the waveguide gradually decreases, the phase constant increases. Fig. 12 illustrates the change in the mode – i.e. from evanescent below cutoff to propagating mode above cutoff.

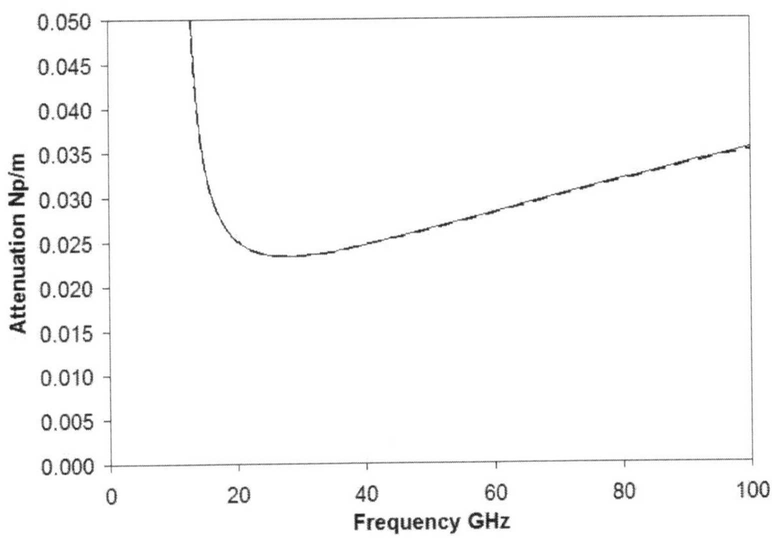

Figure 10. Attenuation of TE_{10} mode from 0 to 100 GHz. the new boundary matching method. power loss method.

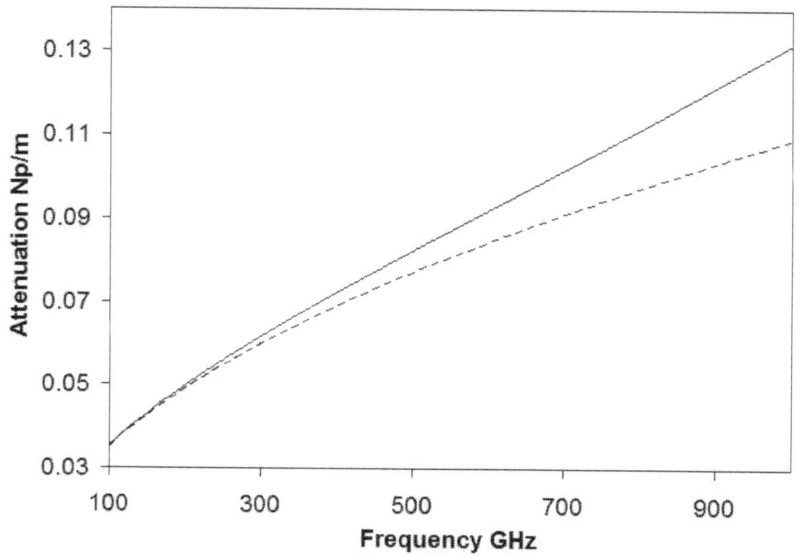

Figure 11. Attenuation of TE_{10} mode from 100 GHz to 1 THz. the new boundary matching method. power loss method.

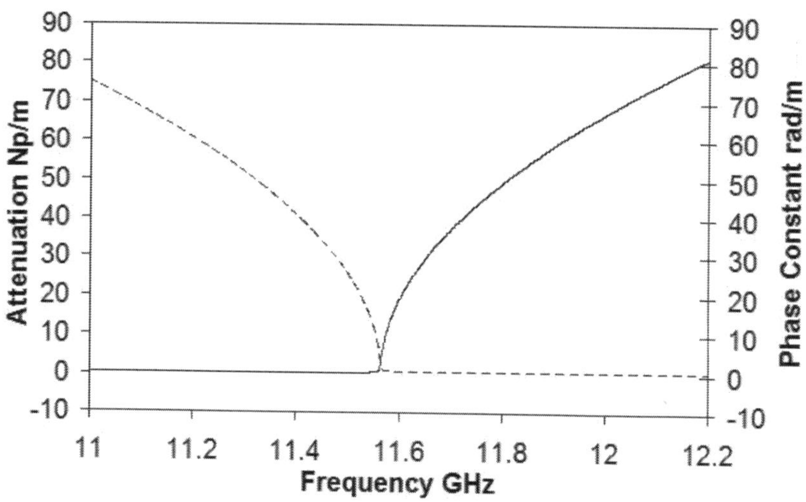

Figure 12. Propagation constant (phase constant and attenuation constant) of TE10 mode in a lossy rectangular waveguide. phase constant. attenuation constant.

7. SUMMARY

A fundamental and accurate technique to compute the propagation constant of waves in a lossy rectangular waveguide is proposed. The formulation is based on matching the fields to the constitutive properties of the material at the boundary. The electromagnetic fields are used in conjunction of the concept of surface impedance to derive transcendental equations, whose roots give values for the wavenumbers in the x and y directions for different TE or TM modes. The wave propagation constant k_z could then be obtained from k_x, k_y, and k_0 using the dispersion relation.

The new boundary-matching method has been validated by comparing the attenuation of the dominant mode with the S21 measurement, as well as, that obtained from the power-loss method. The attenuation curve plotted using the new method matches with the power-loss method at a reasonable range of frequencies above the cutoff. There are however two regions where both curves are found to differ significantly. At frequencies below the cutoff f_c, the power-loss method diverges to infinity with a singularity at frequency $f = f_c$. The new method, however, shows that the signal increases to a highly attenuating mode as the frequencies drop below f_c. Indeed, such result agrees very closely with the measurement result, therefore, verifying the validity of the new method. At frequencies above 100 GHz, the attenuation obtained using the new method increases beyond that predicted by the power-loss method. At f above the millimeter wavelengths, the field in a lossy waveguide can no longer be approximated to those of the lossless case. The additional loss predicted by the new boundary-matching method is attributed to the presence of the longitudinal E_z component in hybrid modes.

REFERENCES

1. A. M. Boifot, E. Lier, T. Schaug-Petersen, 1990Simple and broadband orthomode transducer Proceedings of IEE137 396400

2. H. Booker, 1982 Energy in Electromagnetism. 1st Edition. Peter Peregrinus.

3. M. C. Carter, A. Baryshev, M. Harman, B. Lazareff, J. Lamb, S. Navarro, D. John, A. Fontana, L. , G. Ediss, C. Y. Tham, S. Withington, F. Tercero, R. Nesti, G. Tan, H. , Y. Sekimoto, M. Matsunaga, H. Ogawa, S. Claude, 2004 ALMA front-end optics. Proceedings of the

Society of Photo Optical Instrumentation Engineers, 5489, 10741084 .

4. G. Chattopadhyay, E. Schlecht, F. Maiwald, R. J. Dengler, J. C. Pearson, I. Mehdi, 2002 Frequency multiplier response to spurious signals and its effects on local oscillator systems in millimeter and submillimeter wavelengths. Proceedings of the Society of Photo-Optical Instrumentation Engineers, 4855, 480488 .

5. D. K. Cheng, 1989 Field and Wave Electromagnetics, Addison Wesley, 0-20152-820-7

6. R. C. Chou, S. W. Lee, 1988 Modal attenuation in multilayered coated waveguides. IEEE Transactions on Microwave Theory and Techniques, 36, 11671176 .

7. P. J. B. Claricoats, 1960a Propagation along unbounded and bounded dielectric rods: Part 1. Propagation along an unbounded dielectric rod. IEE Monograph, 409E, 170176 .

8. P. J. B. Claricoats, 1960b Propagation along unbounded and bounded dielectric rods: Part 2. Propagation along a dielectric rod contained in a circular waveguide. IEE Monograph, 409E, 177185 .

9. R. E. Collin, 1991 Field Theory of Guided Waves, John Wiley & Sons, 0-87942-237-8 York.

10. J. I. Glaser, 1969 Attenuation and guidance of modes on hollow dielectric waveguides. IEEE Transactions on Microwave Theory and Techniques (Correspondence), 17, 173176 .

11. C. Kittel, 1986 Introduction to Solid State Physics, John Wiley & Sons, New York.

12. J. , 1998 TM-properties of HTS's rectangular waveguides with Meissner boundary condition. International Journal of Infrared and Millimeter Waves, 19, 399408 .

13. S. Paine, D. C. Papa, R. L. Leombruno, X. Zhang, R. Blundell, 1994 Beam waveguide and receiver optics for the SMA. Proceedings of the 5th International Symposium on Space Terahertz Technology, University of Michigan, Ann Arbor, Michigan.

14. O. M. A. Seida, 2003 Propagation of electromagnetic waves in a rectangular tunnel. Applied Mathematics and Computation, 136, 405413 .

15. N. U. Shankar, 1986 Application of digital techniques to radio astronomy measurements, Ph.D. Thesis. Raman Research Institute. Bangalore University.

16. J. A. Stratton, 1941 Electromagnetic Theory, McGraw-Hill, 070621500 York.

17. V. Vassilev, V. Belitsky, 2001a A new 3-dB power divider for millimetre-wavelengths. IEEE Microwave and Wireless Components Letters, 11, 3032 .

18. V. Vassilev, V. Belitsky, 2001b Design of sideband separation SIS mixer for 3 mm band. Proceedings of the 12th International Symposium on Space Terahertz Technology, Shelter Island, San Diego, California.

19. V. Vassilev, V. Belitsky, C. Risacher, I. Lapkin, A. Pavolotsky, E. Sundin, 2004 Design and characterization of a sideband separating SIS mixer for 85115 GHz. Proceedings of the 15th International Symposium on Space Terahertz Technology, Hotel Northampton, Northampton, Masachusetts.

20. C. K. Walker, J. W. Kooi, M. Chan, H. G. Leduc, P. L. Schaffer, J. E. Carlstrom, T. G. Phillips, 1992 A low-noise 492 GHz SIS waveguide receiver. International Journal of Infrared and Millimeter Waves, 13, 785798 .

21. Y. Wang, Z. A. Qiu, R. Yalamanchili, 1994 Meissner model of superconducting rectangular waveguides. International Journal of Electronics, 76, 11511171 .

22. J. H. Winters, C. Rose, 1991 High-Tc superconductors waveguides: Theory and applications. IEEE Transactions on Microwave Theory and Techniques, 39, 617623 .

23. S. Withington, 2003 Terahertz astronomical telescopes and instrumentation. Philosophical Transactions of the Royal Society of London, 362, 395402 .

24. R. Yalamanchili, Z. A. Qiu, Y. Wang, 1995 Rectangular waveguides with two conventional and two superconducting walls. International Journal of Electronics, 78, 715727 .

25. G. Yassin, C. Y. Tham, S. Withington, 2003 Propagation in lossy and superconducting cylindrical waveguides. Proceedings of the 14th

International Symposium on Space Terahertz Technology, Tucson, Az.

26. K. H. Yeap, C. Y. Tham, K. C. Yeong, 2009a Attenuation of the dominant mode in a lossy rectangular waveguide. Proceedings of the IEEE 9th Malaysia International Conference on Communications, KL., Malaysia.

27. K. H. Yeap, C. Y. Tham, K. C. Yeong, E. H. Lim, 2010 Full wave analysis of normal and superconducting microstrip transmission lines. Frequenz Journal of RF-Engineering and Telecommunications, 64, 5966 .

28. K. H. Yeap, C. Y. Tham, K. C. Yeong, K. H. Yeap, 2009b A simple method for calculating attenuation in waveguides. Frequenz Journal of RF-Engineering and Telecommunications, 63, 236240 .

INDEX